DOUGLAS LAKE RANCH

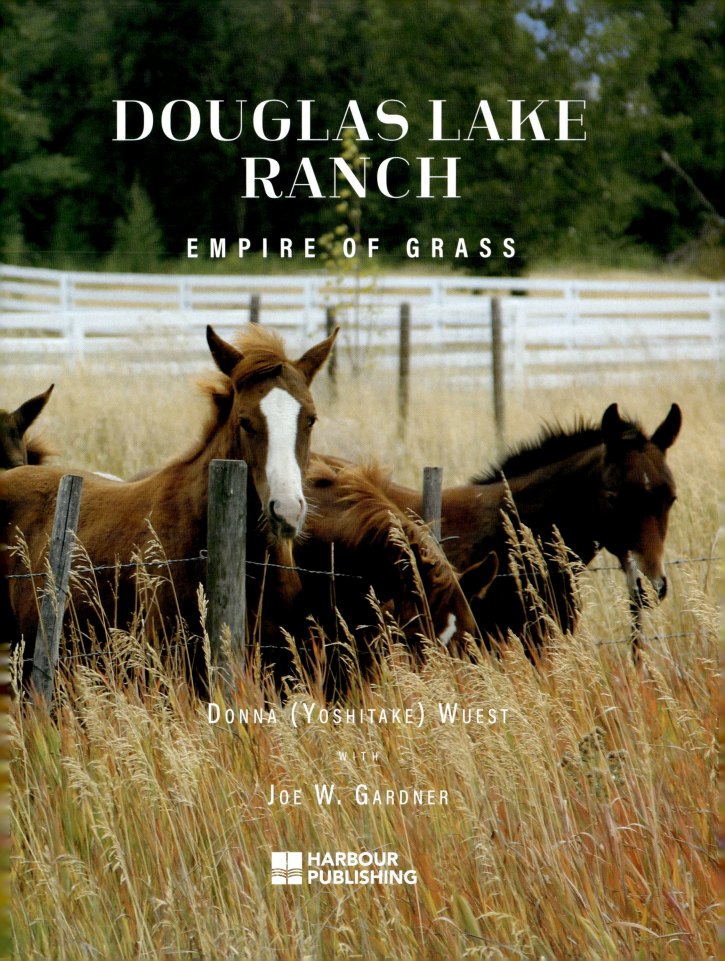

Copyright © 2023 Donna (Yoshitake) Wuest

1 2 3 4 5 — 27 26 25 24 23

All rights reserved. No part of this publication may be reproduced, stored in a retrieval system or transmitted, in any form or by any means, without prior permission of the publisher or, in the case of photocopying or other reprographic copying, a licence from Access Copyright, www.accesscopyright.ca, 1-800-893-5777, info@accesscopyright.ca.

Harbour Publishing Co. Ltd.
P.O. Box 219, Madeira Park, BC, V0N 2H0
www.harbourpublishing.com

Edited by Noel Hudson
Cover and text design by Roger Handling/Terra Firma Digital Arts
All images courtesy Joe W. Gardner unless otherwise indicated
Title page photo by Yuki Sageishi
Cover photo by Edward Hennan
Printed and bound in South Korea

Harbour Publishing acknowledges the support of the Canada Council for the Arts, the Government of Canada and the Province of British Columbia through the BC Arts Council.

Library and Archives Canada Cataloguing in Publication

Title: Douglas Lake Ranch : empire of grass / Donna (Yoshitake) Wuest with Joe W. Gardner.
Names: Wuest, Donna Yoshitake, author. | Gardner, Joe W., author.
Description: Includes bibliographical references and index.
Identifiers: Canadiana (print) 20230503187 | Canadiana (ebook) 2023050325X | ISBN 9781990776427 (hardcover) | ISBN 9781990776434 (EPUB)
Subjects: LCSH: Ranches—British Columbia—History. | LCSH: Ranching—British Columbia—History.
Classification: LCC FC3820.R3 W84 2023 | DDC 971.1/7203—dc23

Contents

Preface	6
Introduction	11
History	**15**
The Woodward Era	**29**
The Joe Gardner Era 1979–2019	**43**
Chunky Woodward Dies	**75**
The Bernard Ebbers/WorldCom Connection	**83**
The Stan Kroenke Era	**88**
Cattle—The Number One Business	**101**
Precious Grasslands	**111**
Farming	**122**
Timber Resources	**132**
Recreation Operations	**135**
The Douglas Lake Ranch Community	**142**
Douglas Lake Neighbours	**164**
The Evolution Continues	**168**
Appendix A — Alkali Lake Ranch	172
Appendix B — Circle S Ranch	174
Appendix C — Quilchena Ranch	177
Appendix D — Riske Creek Ranches	179
Appendix E — Gang Ranch	184
Bibliography	188
Index	189

Preface

I grew up on a large BC ranch that at the time had more apple orchards than cattle ranges. My family worked as labourers in the orchards. I did, as well, during school breaks and summer holidays, as soon as I turned thirteen years old. The work was back-breaking and dirty and carried on regardless of the weather. Those years gave me an appreciation for farmers, ranchers and all their "hands," who work so very hard to grow the products that become our food and who thankfully persist through adversities, most caused by factors over which they have no control. The owners of agricultural lands and enterprises—whether they are owner-operators, multi-generation families, cooperatives, conglomerates, or local or foreigner investors—are the ones who enable the agricultural industry to exist. Thank you to all of them and to the multitude of other people who contribute at each step in the food-growing and production processes. This book recognizes the thousands of people who have owned, managed and worked at Douglas Lake during the ranch's almost century-and-a-half existence, particularly the last half-century.

After writing a book on the history of the Coldstream Ranch, where I grew up, I wanted to write about the largest ranch in Canada, which is also in British Columbia. Douglas Lake encompasses more than a million acres, stretching from the south-central to the central interior of BC. In 2016, when Douglas Lake general manager Joe Gardner and I started talking about updating the documented history of the ranch, he had been in that position for thirty-seven years and was already preparing for his retirement in 2019. Since the previous Douglas Lake history book ended in 1979, the same year Gardner became general manager, the logical sequel would focus on the forty years that he was in that position. And, since Gardner diligently kept detailed daily journals throughout his tenure, this part of the ranch's history could include first-hand anecdotes, which I find the most interesting kind of history to write.

Two previous books, *Three Bar: The Story of Douglas Lake*, by Campbell Carroll, and *Cattle Ranch: The Story of the Douglas Lake Cattle Company*, by Nina Woolliams, provided comprehensive information about the early history of Douglas Lake. This book covers the next forty years, which saw significant changes but also the continuation of important practices for the ranch's ongoing success.

I am grateful for Gardner's many contributions of history, stories and information, which reveal the evolution of the ranch, and for his time and assistance in completing this book. During my research, I also spoke with many other Douglas Lake employees, some of whom have worked at the ranch for as many or even more years than Gardner did, some of whom were relatively recent additions to the ranch community; some are named, and some are not. Their stories give the book richness and depth that is possible only with first-hand accounts. I feel privileged that these people shared with me, and therefore with you, a glimpse of their lives at Douglas Lake. Thank you all!

Thanks to current Douglas Lake general manager Phil Braig for his support of this book. Special thanks to Bobbi Parkes, executive assistant to the Douglas Lake general manager, working with Gardner and then Braig, for her valuable help in arranging meetings, providing contact information, and finding photos for this book. Map wizard Dale Arnell created the maps (see pages 10, 97 and 98), which will help you find the key Douglas Lake locations and give you an idea of the size of the Douglas Lake ranches.

The Woodward family provided wonderful memories of the years when they owned Douglas Lake. Particular thanks to Kip Woodward for coordinating the family's photographs and being supportive of this book.

My husband, Jim, drove me to Douglas Lake and the northern ranches numerous times while I researched and interviewed, and I'm grateful for his encouragement and dependable support of all my writing endeavours.

Once again, the talented and creative team at Harbour Publishing have turned my manuscript and the many photographs into a wonderful history of ranching in BC. Editor Noel Hudson worked diligently to ensure the book read smoothly and coherently, and I am grateful for all his help in making this the best book it could be. I feel honoured and proud to have my book published by this award-winning publisher.

Donna (Yoshitake) Wuest
Langley, BC, 2023

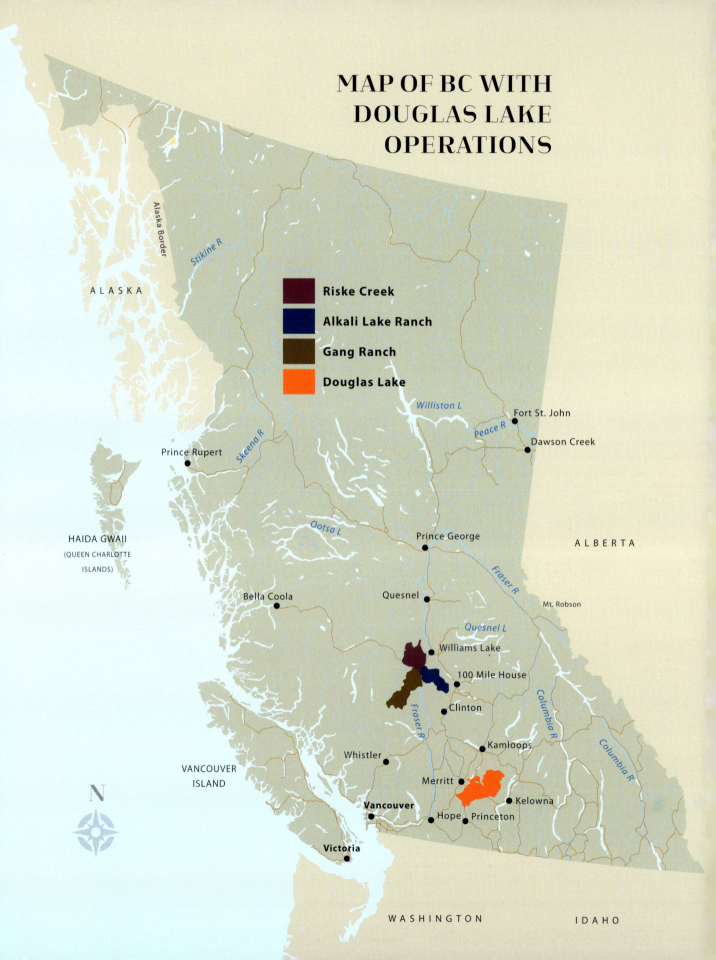

Introduction

According to Statistics Canada, beef consumption per capita in 2021 was 28.1 kilograms (close to 62 pounds). Canadians consume a million tonnes of beef a year. Douglas Lake Cattle Company ranches collectively produce more beef than any other ranching enterprise in Canada. Size matters in any agricultural venture, with economy-of-scale efficiencies and diversification contributing to operational and financial stability, and providing the resilience necessary to navigate unpredictable challenges such as severe weather, wildfires and even a global pandemic.

During the century and a half that Douglas Lake has existed as a ranching enterprise, owners and managers have been passionate about stewardship of their land and sustainability of their natural environment.

An essential component of Douglas Lake's success is the grasslands, which, when effectively managed, provide low-cost feed for cattle. When the earliest European settlers preempted these lands in the 1870s, they recognized the value of the grasslands for raising livestock. By managing the grazing of the grasslands to protect and preserve the indigenous grasses, primarily bluebunch wheatgrass, this valuable natural resource has remained sustainable.

Managing the grasslands sustainably is a science. It involves skills such as fencing to contain the cattle in a particular grazing area for an appropriate length of time, herding the cattle to areas that might not otherwise be grazed, and preventing infestations of noxious invasive weeds and introduction of non-native species by human

Map Dale Arnell / Roger Handling

Previous pages:

Brent Gill

BC Department of Recreation & Conservation historical marker.

activity. Grassland stewardship is the responsibility of everyone who has contact with this precious and sensitive natural resource.

The grasslands are also home to wildlife, and Douglas Lake has identified more than 400 species of flora and fauna—some at-risk and some endangered—that live in the ranches' ecosystems.

From the beginning, Douglas Lake owners have been committed to keeping the ranches intact and, even as expenses have escalated, have had the financial depth to back that commitment. The management and employees of Douglas Lake are dedicated to following through on that commitment.

The gate at the entrance to the Douglas Lake Home Ranch.
Yuki Sageishi

INTRODUCTION

History

IN THE BEGINNING…

The genesis of the Douglas Lake ranchlands was the 320 acres at Round Lake (now called Sanctuary Lake) preempted by Scottish immigrant John Douglas in 1872. The British Columbia Land Ordinance Act of 1870 enabled unoccupied, non-reserved, unsurveyed Crown land to be claimed for settlement and agricultural purposes by staking the land and filing an application with the provincial government. Once Douglas had fulfilled land improvement and survey requirements, he received a Crown grant and became the registered owner of the land, at which time he was eligible to proceed with additional preemptions. Eventually, he assembled 900 acres, on which he raised up to 1,400 head of cattle, making Douglas Lake one of the largest ranches in the Upper Nicola Valley at the time.

Douglas ranged his cattle on Crown land south of his ranch from spring until fall, then moved them back to his home ranch for the winter, feeding them wild hay he had cut during the summer. Having that winter feed was crucial for the cattle's survival during the severe weather that occurs in the area.

Finding accessible markets for the cattle was another major hurdle for not only Douglas but also all other ranchers in the Nicola Valley. The two back-to-back gold rushes that had brought thousands of prospectors to the province were over by the time Douglas started ranching (the Fraser River gold rush started in 1858 and the Cariboo gold rush started two years later,

Opposite: In 1872, John Douglas preempted the 320 acres that became the nucleus of the Douglas Lake ranchlands.
Image A–02111 courtesy Royal BC Museum

but both had ended by the mid-1860s). The demand for beef had shifted from Interior mining towns to settlements in New Westminster and Victoria.

Transporting stock to the coast was the problem. Driving the cattle along treacherous trails with little feed was a horrendous undertaking, but those that reached the Fraser River could be transported by steamer to the coast. While moving cattle via rail would eventually become a shipping solution, construction of the transcontinental railway, a condition of British Columbia joining Confederation in 1871, didn't get underway until 1881.

However, when the Canadian Pacific Railway (CPR) did commence construction in the BC Interior, a new local cattle market emerged. The thousands of labourers required to build the railway through difficult terrain had to be fed, as did the people who were drawn to the new towns that sprang up adjacent to the railway.

GROWING TO MEET A NEED

The CPR called for tenders to supply beef for the construction workers, but the requirements were beyond the capabilities of most individual ranchers such as Douglas. However, for an entrepreneur like Joseph Greaves, who was experienced at raising cattle, selling them and even butchering them, the railway's call was a customized opportunity.

By 1845, when teenager Joseph Blackbourne Greaves arrived in New York from his home in England, he'd already learned the skills of a butcher from his father, earned money betting on horses, and finagled his way across the Atlantic by looking after pigs. He was one of the hundreds of thousands of emigrants from England who, during the mid-1800s, endured the grueling two-week sea voyage to seek their fortunes in America. After landing on the east coast, Greaves worked his way across the country. In California, he worked as a butcher before heading to Oregon, where he drove cattle and sheep north to the Cariboo gold fields. He sold the livestock, again worked as a butcher, then returned to Oregon for more cattle. By the time he arrived back in the Cariboo, though, the gold rush was over, so Greaves and his cattle headed south again. Eventually, he settled on the south banks of the Thompson River, where he bought land and raised cattle to sell in Victoria.

When Greaves heard that the CPR was looking for a guaranteed supply of fresh beef for work crews, he approached butchers and ranchers he had met throughout BC, hoping to assemble the capability to pursue the contract. Greaves's contacts reached out to their contacts, and in short order he had access to the supply necessary to achieve his objective of controlling the BC beef market and securing the CPR contract. Next, Greaves created a syndicate consisting of:

Joseph Blackbourne Greaves, the entrepreneur.
Courtesy Nicola Valley Museum & Archives

- Benjamin Van Volkenburgh, who owned a Victoria meat market, a wharf to unload livestock, a fenced grazing area for the stock prior to slaughter, and an abattoir
- Joseph Despard Pemberton, who had been a surveyor, a member of the Legislative Assembly, and who raised cattle and horses
- William Curtis Ward, a banker with ten children, a fact that would become significant in the future of the Douglas Lake Cattle Company
- Charles William Ringler Thomson, who had been a commercial fisher, farmer and the manager of Victoria Gas Company
- Peter O'Reilly, a former gold commissioner, magistrate and county court judge

Greaves and his five partners formalized their syndicate in 1882, with Greaves as manager. His strategy was to buy cattle from ranchers throughout the BC Interior with a cash deposit and mutually agreeable terms for the

William Curtis Ward, a member of Joseph Greaves's syndicate.
Courtesy Nicola Valley Museum & Archives

balance. The cattle were to be pastured on the sellers' ranches until needed. Despite losing the railway contract to Thaddeus Harper, who at the time had large herds of cattle just east of Kamloops, Greaves's syndicate did well the first year, selling to markets on the coast. By the following year, Harper was unable to supply the beef as contracted because the Greaves syndicate was buying all the available cattle. When Harper defaulted in 1883, the syndicate took over the CPR contract, delivering 300 head of cattle a week to the point of construction.

Earlier that same year, Greaves had met Charles Beak, who was by then owner of one of the largest ranching operations in the Nicola Valley. Like Greaves, Charles Miles Beak had driven cattle and sheep from Oregon to the Cariboo during the gold rush. Beak had settled in the area to raise beef and dairy cattle and opened a butcher shop in Barkerville. In the late 1870s, he had moved to the Nicola Valley, where he'd recognized the ranching potential of the bunchgrass hillsides. He returned to England to raise the money necessary for the large ranching operation he envisioned, and on his return, in 1883, started a buying spree. One of his first acquisition targets was John Douglas's ranch, as by then Douglas was ready to sell his land and cattle. With Douglas's ranch as the core property, Beak continued to buy other ranches until he owned a significant portion of the land around the north and east sides of Douglas Lake. When he bought the ranch on the northeast shore of Chapperon Lake, he became one of the largest ranchers in the Interior.

BEAK AND GREAVES became partners, with each owning an undivided half of the more than 8,000 acres of land, 5,000 head of cattle, and 60 horses that Beak had purchased. Greaves added 1,000 of his own cattle to the herd. Beak continued to manage the ranch and acquire more ranches in the Douglas Lake area, while Greaves managed the syndicate, which had succeeded in cornering the beef market.

In 1884, with their initial investment quadrupled, four of Greaves's syndicate members—Van Volkenburgh, Pemberton, Ward and O'Reilly—sold their shares to Thomson, Beak and Greaves. By then, the syndicate owned more than 8,500 head of stock, some at ranches where Greaves had initially purchased them and the rest on the 8,000 acres of deeded land that Greaves and Beak owned around Douglas Lake.

DOUGLAS LAKE RANCH BEGINS

Greaves and Beak invited Thomson and Ward to join their ranching operation, and on June 4, 1884, they formed Douglas Lake Ranch, with Beak as interim manager and Greaves continuing the operations of the syndicate. Ward bought back into the syndicate, and the four became equal partners in both operations. The "111" brand became Douglas Lake's official brand.

Beak focused on buying land at lower elevations capable of growing sufficient winter feed for the livestock. Cattle were ranged on Crown lands until frost and snow arrived on those higher elevations, at which time they were moved down to where bunchgrass on deeded land provided winter forage.

By 1885, as construction of the railway neared completion and that market for beef was coming to an end, Beak moved to Victoria to open a butcher

Joseph Greaves managed Douglas Lake from 1885 to 1910.
Courtesy Nicola Valley Museum & Archives

outlet for Douglas Lake cattle. The ranch also sold to other butchers on the coast, shipping livestock via the new railway as it opened. Greaves took over management of Douglas Lake and continued to buy land, including more than 14,000 acres around Chapperon Lake.

On June 30, 1886, Greaves, Beak, Thomson and Ward incorporated The Douglas Lake Cattle Company, Limited Liability, to raise, buy and sell cattle and horses, buy and sell land, and carry on farming and butchering businesses. By then, the ranch owned 22,765 acres and 12,000 head of cattle, with each of the partners having a quarter interest.

Greaves continued to buy land and cattle despite the disastrous weather conditions that had resulted in a shortage of winter feed and the death of thousands of cattle during the first year of ranching operations for the new company. That tough winter convinced Greaves he needed to improve the hardiness of the Douglas Lake stock. He brought some of the first Herefords to the Nicola Valley and bred them with existing Shorthorns to develop a crossbred strain that he thought would fare better in the tough winters ahead, with the added advantage of hybrid vigour. Before long, he had increased the herd size to 18,000 head.

Beak, meanwhile, turned his attention to Clydesdale horses. He returned from a trip to Britain with a prize-winning two-year-old stallion called "The Boss," which he had purchased in Glasgow, shipped across the Atlantic, and then across Canada to Douglas Lake—a long and arduous trip for both horse and new owner. "The Boss" is registered in the Clydesdale Stud Book of Canada and became one of the foundations of the breed in this country.

Greaves was angry with Beak's investment but nevertheless agreed to the purchase of six registered Clydesdale mares from Ontario. The investments turned out to be more profitable than either had anticipated and, for years, sales of the heavy draft-horse colts covered most of the ranch's expenses.

In 1889, grasshoppers attacked the fields in the Nicola Valley, but the disastrous event turned out to be another opportunity for Greaves, who bought ranches from owners who could not withstand the scourge. The following year, he acquired grazing rights on another 10,000 acres of Crown land.

By 1892, Douglas Lake Ranch had doubled in size to more than 47,000 deeded acres. Greaves and Beak disagreed on the direction of further land purchases, and when their disagreements spread to virtually all decisions, Beak wanted out. He sold his shares to the other three owners.

The next bad winter killed more than 8,000 cattle, and Greaves recognized that he had to improve hay-growing capability for winter feed. He dammed Chapperon Creek and built a sluice gate at the outlet to create a water reservoir to irrigate the fields.

With steady horse and cattle sales, improved land productivity and good range management, Douglas Lake prospered. More than a hundred people worked on the ranch, not only as part of the haying and cowboying crews, but also fencing crews, irrigation crews, teamsters, carpenters, gardeners to grow vegetables, and farmers to raise chickens. The ranch had a busy store, a saddlery, a cookhouse and everything else needed to provide for the people who lived and worked there. Greaves created a sense of community, with family-style dinners and Sunday church services. He financed local ranchers who pledged beef for fall delivery, and he participated actively in agricultural associations.

RANCH FOR SALE

By the early 1900s, as the owners aged and with Greaves edging toward retirement, selling the ranch became a priority. A number of prospective buyers were interested, including some prominent individuals, but a ranch

In 1909, at the age of seventy-nine, Joseph Greaves was ready to retire. *Courtesy Nicola Valley Museum & Archives*

of this size was difficult to sell. Despite ongoing challenges and uncertainty due to extreme weather and labour shortages, Greaves continued to take advantage of every opportunity to buy more land and cattle.

By 1905, selling the ranch became even more urgent for its aging owners. Thomson's valuation of the ranch, including the almost 100,000 acres of land, 14,000 head of cattle and horses, machinery, 300 miles of fencing, and stock in the store, was more than a million dollars. Efforts to sell—even attempts to raise funds on the London Stock Exchange—failed. The ranch was taken off the market, but the partners continued to search unsuccessfully for a buyer.

Frank Bulkley Ward, reluctant rancher, preferred to play polo. Frank is second from left in this photograph of the 1913 Nicola Polo Team. *Courtesy Nicola Valley Museum & Archives*

Frank Bulkley Ward was born in Victoria, one of William and Lydia's ten children. At age sixteen, he joined his father's bank, but when his doctor recommended that he work outdoors because of a chronic chest condition, Frank went to Douglas Lake, where his father was a part-owner. During the year that he worked there as a cowboy, he wasn't well accepted. Unlike the rest of the cattle crew, he rode English style and dressed in more aristocratic attire. After returning to the bank for a brief period, he joined his father's friend Joseph Pemberton in buying a ranch in southern Alberta. Though Frank managed that ranch, he preferred to train and sell polo ponies and to play polo in Calgary. Unable to expand their Alberta ranch, Pemberton and Ward chose to sell and return to BC. That's when William Ward started urging his partners at Douglas Lake to take Frank on as assistant manager, which finally happened just prior to William becoming the sole owner of the ranch.

HISTORY

William Ward pushed for his son Frank to become Greaves's assistant manager. While Greaves refused at first, he finally relented in October 1909. A year later, at age seventy-nine, Greaves retired and sold his shares in the company. Thomson wanted to sell as well, and on August 3, 1910, The Douglas Lake Cattle Company, Limited Liability, emerged with William Ward as sole shareholder and Frank as ranch manager. The ranch consisted of more than 100,000 acres of deeded land, on which 10,000 cattle wintered.

William Ward continued to look for potential buyers for Douglas Lake, and when he was unsuccessful, he decided to keep the operation as a family investment. In June 1914, Douglas Lake Cattle Company became a private Ward family company, with each of the ten children owning equal shares and Frank continuing as manager.

A month after the Ward family assumed ownership of the ranch, World War I broke out. While cattle prices soared, so too did operating costs and challenges. As half a million Canadian men went to war, only those who were too old or unfit for military service remained at home. The availability of labourers to work on the ranch became a serious problem, and because of the labour shortage, wages increased.

The need for more grazing land was emphasized after another infestation of grasshoppers and nuisance caused by an abundance of beavers blocking irrigation ditches. Even with the purchase of more grazing land, the ranch still had to decrease its herd size to avoid permanent damage from overgrazing.

By the end of World War I, Douglas Lake had grown to 121,000 acres, but the herd size was down to 8,200, and while beef prices were rising, they were still lower than peak wartime prices. Although the ranch was not on the market, a prospective American buyer offered the Ward family $1.5 million. When the family pushed for $1.75 million, they lost the deal.

FROM BAD TO WORSE

By the summer of 1921, cattle prices had dropped to half of peak prices during the War, the grasshoppers returned, labour was still difficult to attract and retain, and Douglas Lake's herd continued to shrink.

After William Ward died in 1922, the American buyer who had previously been interested in buying the ranch offered the Ward family $1.3 million. While Frank was willing to sell, his siblings rejected the offer, insisting on $1.5 million. Again, the sale was lost.

By 1927 the ranch's cattle herd had dropped to 75 per cent of its size a decade earlier, and as mechanization decreased the demand for horses, their numbers were reduced by half as well. Grasshoppers and forest fires caused more crop and stock losses, and demand and prices for cattle continued to drop. Frank tried unsuccessfully to sell the ranch for $800,000, and while a wealthy New Yorker was interested in the property for recreation purposes, even at $750,000 that sale never happened.

A year later, the ranges improved, arsenic was used as a temporary solution for the grasshopper problem, beef prices increased, and Frank began to rebuild the herd size. However, with the Wall Street crash in late 1929, efforts to sell the ranch were again suspended. Douglas Lake Cattle Company was restructured from a private family company to a family corporation to reduce federal income taxes.

As the Great Depression took hold in the early 1930s, beef prices hit yet another all-time low, but Frank continued to rebuild the herd size and in 1933 finally exceeded his goal of 10,000 cattle.

The following year, forest fires burned 300 miles of ranch fences, which was a devastating loss at a time when expenses were too high and beef prices too low. However, the fences had to be rebuilt—at a cost of $250 per mile.

The ranch once again went on the market. Offers to buy the ranch dropped to as low as $300,000, and one Vancouver realtor even suggested the Douglas Lake Cattle Company be exchanged for a nine-storey building in St. Paul, Minnesota.

The Ward shareholders rejected a suggestion by the trustees and directors of the company that the ranch be put on the market at $600,000. The siblings wanted to reduce overhead expenses to try to improve the selling price. In 1937, when his shareholder siblings suggested that Frank retire, a new manager be appointed, and all the expensive foremen be fired, Frank angrily refused. However, after another severe winter, Frank had had enough. On April 30, 1940, at the age of sixty-five, Frank Ward retired, and Brian Chance took over as manager.

Born in Australia and educated in England, Brian Kestevan de Peyster Chance had spent four months at Douglas Lake in 1921 before returning to Australia to work at sheep stations for five years. He had accepted a job offer to return to Douglas Lake in 1926.

When he became manager fourteen years later, the 142,770 acres of deeded land included fields that could produce enough winter feed for the 10,000 cattle. The ranch, though, was in the midst of another round of difficult years.

Brian Chance, Douglas Lake's first non-owner manager. *Courtesy Nicola Valley Museum & Archives*

HISTORY 25

World War II caused further labour shortages, and government wartime regulations severely affected all ranchers. The federal government imposed an embargo on cattle exports to the US yet set a low ceiling price on beef at home and rationed it to two pounds a week per person.

In the mid-1940s, one of the worst grasshopper outbreaks in BC history hit the Nicola Valley and coincided with a shortage of water for irrigation. The hay harvest didn't meet Douglas Lake's winter feed requirements, and hay and grain had to be purchased from the Prairies.

In 1947, the first spraying for grasshoppers by airplane ended the infestation. That fall and winter, the ranch used aircraft for the first time to locate the cattle that inevitably evaded the cowboys.

NEW OWNERS

Douglas Lake Cattle Company had been off the market during World War II, but after the War ended, the Ward family resumed efforts to sell. Chance suggested a price no less than $1.5 million.

In August 1950, Colonel Victor Spencer and W.P. (Bill) Studdert purchased the 143,250-acre ranch for $1.4 million.

Victor Spencer's family owned David Spencer Ltd., a chain of dry goods stores, and though he had joined the family business and worked there at various times, he was more interested in his in-laws' ranching business. After returning from the Boer War and, later, World War I, he had retired as a lieutenant-colonel and thereafter was known as Colonel Spencer. Spencer married Gertrude Winch and acquired his father-in-law's Earlscourt Ranch in Lytton. He also acquired Bryson's Ranch, Circle S Ranch at Dog Creek, and Diamond S Ranch at Pavilion before purchasing Douglas Lake Cattle Company.

Bill Studdert started as a deck-hand on fish boats, became a certified sea captain, and owned fish canneries and eventually two freighters. He also owned T Bar 3 Ranch in Philipsburg, Montana, and in 1948, along with Floyd Skelton, bought the Gang Ranch, in BC's Chilcotin region, before joining Spencer in purchasing Douglas Lake.

For a year, Spencer was president, Studdert was managing director and vice-president, and Chance remained as manager of Douglas Lake. The two owners invited Frank Mackenzie Ross to join them in their venture, and for a brief period of time the three were equal one-third owners. However, Ross decided he did not want to be involved in the business with Studdert, so Spencer negotiated to buy back Studdert's shares. On April 25, 1951, Spencer and Ross became equal owners of a new Douglas Lake Cattle Company.

Frank Mackenzie Ross was born in Scotland, fought in World War I, and was awarded the Military Cross. During World War II, he worked in Ottawa procuring supplies for the British Admiralty, which earned him the British Companion of the Order of St. Michael and St. George (CMG). He had a successful career in finance and industry and married the highly accomplished economist Phyllis Turner, whose son John became Prime Minister of Canada for a brief period in 1984. From 1955 to 1960, Ross was the 19th Lieutenant Governor of British Columbia.

The new owners' demands on ranch management were difficult and perhaps somewhat unreasonable. To pay for the purchase of the ranch, the owners increased annual cattle sales from the usual 25 per cent to 50 per cent of the herd. That meant yearling heifers had to deliver calves as two-year-olds, and as a result many of the calves and some of the young mothers died. When the owners' demands escalated to selling yearlings as well as two-year-olds, the entire ranching operation was jeopardized. Expenditures were curtailed so severely that no new machinery, fences or irrigation infrastructure could be purchased. Yet the owners purchased more land.

By 1957 Spencer and Ross were disenchanted with the cattle business. Profits were poor, and they felt that too many variables were out of their control. However, despite their aversion to investing in their operations, the ranch almost doubled in price.

On July 20, 1959, Spencer and Ross completed the sale of Douglas Lake Cattle Company to Charles "Chunky" Woodward and his investment banker friend John West for $2.6 million.

The Woodward Era

The purchase of Douglas Lake Cattle Company by Charles "Chunky" Woodward and John West in 1959 began thirty-five years and two generations of ownership by the Woodward family.

Charles Namby Wynn "Chunky" Woodward was the grandson of Charles A. Woodward, founder of Woodward's Stores Limited, and the son of William C. Woodward, Lieutenant Governor of BC between 1941 and 1946. At age thirty-two, "Chunky" became president of the department store chain, but his heart was in ranching, having spent his childhood visiting Alkali Lake Ranch, in the Cariboo, which was owned by his maternal grandfather, Charles E. Wynn Johnson, a close friend of Frank Ward.

"Ranching was an important part of Chunky's lifestyle," says his widow, Carol. "He'd shed his city suits, put on jeans and a cowboy hat, and become part of the ranch. He was passionate about the ranch. He loved the ranch. We'd go up on weekends, particularly during hunting season. He loved the outdoors. The ranch suited him."

Chunky Woodward's close friend John Joseph West, a fifth-generation Canadian, was a commerce grad, worked in the securities business, and was an avid sportsman. "Chunky and Dad shared passions for hunting and fishing," says son John West. "Both families had summer places at Savary Island, where we spent a lot of time together fishing with our fathers. The two families shared a house at Douglas Lake, but we were rarely there together because the Woodwards spent the end of each summer there, and our family went up only in the fall, winter or spring, but we would on occasion overlap the use of the house."

Opposite:
Photographed here in 1958, Douglas Lake Ranch was on the cusp of a new era of leadership.
Image I–22386 courtesy Royal BC Museum

Above left: "Chunky" Woodward, a rancher at heart.
Courtesy Kip Woodward

Above right: John West, Woodward's friend and business associate, became part-owner of Douglas Lake.
Courtesy John West Jr.

Woodward and West were long-time friends and business associates. West was a director of Woodward's stores, and the two of them got together regularly. "From what I understand, when Dad and Chunky heard that Spencer and Ross were interested in selling Douglas Lake Cattle Company, Dad looked at the numbers and told Chunky this could be a viable operation once some capital was invested," says John. "Dad didn't think they would lose money. Apparently Chunky told him, if that's what he was advising, he should put his money where his mouth was and come in on the deal, too."

"Spencer and Ross presented the ranch as a unique opportunity," adds Woodward's older son, John. "My father grew up spending time at his grandfather's Alkali Lake Ranch and would much rather have been a cowboy than in retail."

The agreement between Woodward and West was that if either of them died, the other would automatically buy all his partner's shares to become sole owner.

Owner John West and manager Brian Chance.
Courtesy John West Jr.

Early Memories of Woodward and West Family Members

"Douglas Lake came into our family's life when I was ten, and it was marvelous," remembers Ann West. "We would drive up, usually the Hope–Princeton Highway, and after the first trip, we knew we were near once we crossed the first cattle-guard. The Home Ranch was the central point, with a most comfortable house, but more exotically, barns and horses and a bunkhouse and a cookhouse and cowboys. And the small lake out front would freeze for skating in the winter. Manager Brian Chance, foreman Mike, and Scotty the cowboy—I am not sure if we were a momentary diversion to them or a nightmare, but my brother and I loved being there, riding, occasional cattle drives, calving, branding, Quarter Horses, and the incredible sense of space and room."

Ann and her mother, June West.
Courtesy Ann West

WYNN, THE OLDEST of the four Woodward siblings, remembers those day-long drives along the Hope–Princeton Highway to the ranch, too: "First time I went up, I was eight or nine. We went up in the summer, and at Easter, Thanksgiving, Christmas and New Year's. I liked being up there, being out in the open. As kids we weren't allowed to be indoors, even at home, so we were always outside doing something. Kip and I would go exploring, and John and I would go out with lunches in our saddle bags and ride with Scotty."

The Woodward families continued to go to the ranch at Thanksgiving for years. "I made a point of driving back to the ranch from my home in California," says Wynn. "It was a chance for all of us to get together there."

"WHEN I WAS in school in Vancouver, my father would pull me out on a Thursday to go to the ranch," says John, the second oldest of the Woodward siblings. "I was missing so much school that my grades suffered. Later, Kip and I went to Brentwood College on Vancouver Island. I spent three summers riding with the cowboys at the ranch. It was quite the experience."

"THE FIRST TIME I went up to the ranch was in 1960," remembers John West. "We'd go in the fall, maybe at Christmas, Easter, and a few other occasions, usually for duck or goose hunting.

"I worked at the Portland cow camp in the summer of 1966, and two summers in 1967 and 1968 at the Dry Farm. Most of what we did was push the cattle up onto the higher ranges. We'd leave at four in the morning, go back to camp at lunch, have an hour siesta, then go back to work. Some would doctor animals, and the rest of us would go fencing or build whatever had to be built, or split wood for winter. We worked six days a week, and on Saturdays had a weekly shower before heading into Merritt in two or three vehicles. I would sit in the back of the truck, and the others would bring beer out for me. After dinner we'd go to the drive-in theatre, which started at ten. They were always western movies. Then we'd head back after that.

"Maybe the only reason I was hired was because Dad was an owner, but I also knew how to ride. Working on the ranch was one of the best things I ever did, and I thoroughly enjoyed it. The people were great."

The long journey from the city to ranch lands can be beautiful. This is a scenic view of Highway 1, between Spences Bridge and Cache Creek. Highway 1 is the road many modern visitors drive along to visit the ranch's northern operations. *Ferenc / Adobe Stock photo*

THE WOODWARD ERA

John West, as a teenager, working as a cowboy at Douglas Lake.
Courtesy John West

Kip, the youngest Woodward sibling, at about thirteen years of age, after his first successful deer hunting experience with his father.
Courtesy Kip Woodward

Kip and Robyn on the buckboard wagon with Joe, the garbage collector, pulled by Sandy, the white-bearded draft horse, circa early 1960s. *Courtesy Kip Woodward*

"AT THAT TIME, the drive from Vancouver took six and a half hours," Robyn, the second youngest of the four Woodward siblings, remembers. "Sometimes my father would pull a trailer with our ponies, or sometimes a cowboy from the ranch would come with a trailer and haul the horses for us. We'd get to ride in the truck with the cowboy."

Robyn was just two or three years old when she first started going to the ranch. "My brother Kip and I were too young to ride with the older kids, so we spent the mornings with Joe, the garbage man, who drove a buckboard wagon with a big trash bin on the back and went from house to house. The wagon was pulled by Sandy, a big draft horse with a long white beard. We picked up the garbage, then stopped at the pig pen to dump the compost for the pigs, then to the garbage dump. That took a few hours, and we had a great time.

"As we got older, we would run around with cow boss Mike Ferguson's youngest daughter, Debbie, and explore everywhere on the ranch. One summer, Debbie dared us to steal the peas and tomatoes from the vegetable garden. We'd sneak in under the rhubarb leaves and try to steal the vegetables, but the gardener would chase us away with his hoe.

"Riding wasn't an optional activity in our family. Once we were five or six, we could go on long rides on our pony or on a ranch horse. We'd be in the saddle for hours, sometimes just riding through the countryside, sometimes following

Robyn assisting at the annual social branding of calves. "Friends of the ranch who seldom dealt with cattle were invited to participate along with the farm crew," remembers Kip. "The cowboys always had great fun teasing them." *Courtesy Woodward family*

behind a cattle drive. We got up early and went with my father. It was quality time spent with him.

"Once Dad got a plane, we'd go up to the ranch that way, and there was always a cribbage game on the one-hour flight. Being able to get out of the city and be totally away was great. It gave us a sense of independence. My aunt [Twigg White] had a cabin at Salmon Lake, and my cousins and I would ride all day. My aunt didn't like us underfoot, so she'd pack sandwiches and fruit for us and tell us to be home by five. She said if we opened a gate to be sure we closed it. Otherwise, there were no restrictions. It gave us a sense of total freedom. We got to run wild in the massive open spaces.

"Saturday nights, we went to the machine shop, and my father showed movies, mostly about World War II or cowboys and Indians. Our cowboys and local Indigenous people would both come, and they'd cheer for opposite sides."

CHUNKY AND HIS younger sister, Mary "Twigg" White, both loved the times growing up at their grandfather's Alkali Lake Ranch. When Chunky bought Douglas Lake, he leased land to Twigg to build a summer home at Salmon Lake. "My parents leased land there in about 1960 and built a house using a Scandinavian architect," remembers Rhegan, one of five children—Andrew, Melanie, Susan, Rhegan and Joel. "It was a summer house and we had stables on about 300 acres. We had seven or eight horses that we used to ride." When the White family returned the lease to Douglas Lake in the early 2000s, the house was renovated and is now called 'Twigg's Place.'"

The White family would spend every summer at their Salmon Lake property. "The day after school was finished, we'd pack up and go to the ranch until the Labour Day weekend," remembers Rhegan. "Often we'd go at Christmas, Thanksgiving and Easter, too. In total, we'd be there two-and-a-half to three months a year. We got to know the ranch kids and travelled around on horseback." Fifty years later, Rhegan and her family still go to the ranch every year or so.

"My number one memory was going on my first cattle drive, when I was about ten or eleven," says Rhegan. "We met at the Home Ranch at four in the morning and rode for ten to twelve hours. Being a part of that was a big deal and is seared into my memory. I worked as a cowboy one summer. It was a unique life that we grew up in. We were given freedom, no strings. It was phenomenal. I was lucky."

"I WAS SEVEN the first year we went up, and I've been going there ever since," says Melanie. "Rhegan and I, of our siblings, were the two most connected to the ranch. We rode a lot as kids. Those were the times when we were most like a family, Mom and the kids. She would send us out in the morning with a knapsack and tell us to go for the day and return for dinner. It was an amazing childhood.

"One ride we always did was to Joe McCauley's, near Salmon Lake. We used to feel as if we could still feel them there. Another landmark for us was an old log house, Liza's house. Sometimes we would ride past there and knock on her door to ask for some water to drink. To me, she represented a kind of renegade pioneer woman with a kind heart. She was a strong, independent woman.

"Chunky and my mother died within weeks of each other. Mom died just before Easter, and we all went to the ranch for an Easter celebration to remember her. Chunky and I went for a ride around the lake to an area he hadn't been, and that was the last time we were together. He died two weeks later."

Rhegan and "Twigg" White riding at Index Lake, 1984.

THE WOODWARD ERA

For Chunky Woodward, the ranch was an escape from his corporate world into the outdoors he loved, where he could be a rancher, hunt, fish, and ride and raise horses.

While horses had always been an important part of the Douglas Lake ranching operations, in 1963 Woodward launched an expanded Quarter Horse breeding and training program. He purchased the stallion "Peppy San," built a barn, and hired a barn boss and horse trainer. Woodward believed the breed was ideal because of its inherent "cow sense," ability to cut cattle from a herd, and its bursts of speed.

HIS ROYAL HIGHNESS, the late Prince Philip, Duke of Edinburgh, visited Douglas Lake for a few days of relaxation during his 1962 tour of Canada.

Upon seeing a demonstration of the Quarter Horses' skills, the Prince invited Chunky Woodward and some of the members of the Canadian Cutting Horse Association to attend the Royal Windsor Horse Show in 1964. Woodward, riders and horses flew to Britain for a three-month tour that culminated at the Windsor show, where Woodward presented the Prince with the Douglas Lake Quarter Horse stallion that became Prince Philip's and Prince Charles's (now King Charles III's) polo horse. "The stallion was used for breeding a lot of the Prince's polo ponies," says John Woodward.

Years later, John and his wife were invited to a Buckingham Palace event hosted by Prince Philip. "We were coming to the door as Prince Philip arrived," John remembers. "We were introduced, and he wanted to talk about Douglas Lake. He had twenty minutes to greet the guests and spent all twenty talking to me about the ranch."

In 1967, "Peppy San" became the World Champion Cutting Horse, as well as winning titles and trophies across Canada and the United States. The famous Quarter Horse bred mares for a substantial fee and produced superior ranch horses. *Courtesy Woodward family*

"WHEN CHUNKY AND Dad bought the ranch, it was old-school manual labour with very little invested in infrastructure," says John West. "For the first ten to fifteen years or so, they, and then Chunky—after my father died—invested in infrastructure to improve the living conditions for employees and begin automation of many functions."

The introduction of big, efficient machinery to harvest, bale and stack hay crops eliminated some jobs and changed others. When rail service from

Prince Philip and Brian Chance riding at Douglas Lake. *Courtesy Nicola Valley Museum & Archives*

Chunky Woodward chatting with Prince Philip during the Royal Windsor Horse Show in 1964. *Courtesy Woodward family*

Nicola to the coast ceased, it became necessary to transport the cattle by truck to feedlots on the Prairies. In the spring of 1967, Douglas Lake hit a new record of branding 5,000 calves.

That year, after forty-two years of working at Douglas Lake and twenty-seven years as manager, Brian Chance retired, leaving management to Neil Woolliams. By then, the ranch consisted of 163,000 acres of deeded land.

In the summer of 1953, when Neil Woolliams was fifteen years old, he worked as part of the haying crew at Douglas Lake. In 1961, after completing his degree in Commerce and Business Administration, he was invited by Brian Chance to join the ranch full-time with the possibility of becoming assistant manager. He spent three years learning all aspects of the ranch, then was promoted to assistant manager and, in July 1967, to manager.

To continue to produce and market more beef, Woolliams increased the herd size and hay production, and sought ways to improve efficiency. He purchased horse trailers for the cowboys and introduced radiophone communications. He also looked at the ranch's natural resources with a view to generating new revenue by diversifying the operations—selling more timber, harvesting Christmas trees, and starting to develop fishing facilities.

In February 1969, John West Sr. died of pancreatic cancer at the young age of fifty-three. Woodward became the sole owner of Douglas Lake.

"I was incredibly fortunate," says son John West. "After Dad died, Chunky bent over backwards to keep me connected with the ranch. He became like a surrogate father. His son John and I have always been good friends, and Chunky would include me in fishing and hunting trips. I remain close friends with all the Woodward children, and our family often spends Thanksgiving weekend with them when we rent Twigg's Place."

THE RANCH IS blessed with an abundance of lakes that supply water for irrigation, much needed in the semi-arid climate, but for sufficient water to reach the hayfields at the right time, Woodward needed to invest significant funds in irrigation equipment. By the mid-1970s, Douglas Lake had one of the most efficient pressure irrigation systems in the province. With the addition of more efficient haying machinery, the ranch reached all-time production levels.

Cattle marketing changed in the 1970s, as well. Instead of trucking all the cattle to the sale yard in Kamloops, the larger ranches with larger herds, facilities and scales sorted the cattle up at home. The auctioneer and cattle buyers would then arrive, and the cattle would be sold. This became a very popular social event that provided the ranch with better delivery conditions: more pounds sold and less stress on the cattle.

The mid-'70s also brought droughts, severe winter weather, an outbreak of calf scour, depressed cattle prices, bear and cougar attacks, and increasing problems and negative impacts on cattle caused by recreational users trespassing on ranch property.

By 1978, with a new pivot irrigation system, a bigger calf crop than ever, improved prices, and more sophisticated breeding programs, including semen testing and scrotal measurement, the ranch was once again in an upward cycle.

After twenty years at Douglas Lake, twelve years as manager, Neil Woolliams left when he and his family purchased their own cattle and sheep ranch in Australia.

The Joe Gardner Era 1979–2019

Joe Gardner was born in Cornwall, Ontario, in January 1946. Six months later, his family moved to Vancouver, where his father, Dr. Joseph A.F. Gardner, CM, became a chemist with Federal Forest Products Laboratory located at the University of British Columbia (UBC). Joseph Gardner later became Dean of Forestry at UBC.

Growing up, Joe Gardner had after-school and weekend jobs, first cutting lawns, then working in Vancouver's Southlands equestrian stables, mucking stalls and working around horses. In the early 1960s, Gardner told his father that he wanted to get a summer job on a ranch. His father informed him that he didn't know any ranchers, so Joe went straight to the library to look up "ranches in BC." He then wrote a letter to the Douglas Lake Cattle Company.

"My mother was concerned when she heard what I'd done," Gardner remembers, "but our next-door neighbour, Fin Anthony's mother [the late Fin Anthony was the face and voice of the popular Woodward's Stores television commercials] told my mother not to worry because no ranch would hire some kid from the city." Mrs. Anthony was wrong. Gardner got a positive response and soon started the first of three summers working at Douglas Lake.

FIRST DAYS OF WORK

When his mother and father dropped Joe off at Douglas Lake on July 1, 1963, they were greeted by manager Brian Chance, who assured Joe's mother that her son would be okay. After Joe's parents left, Chance pointed out the bunkhouse where Joe would be rooming and informed him that he would be around the Home Ranch for a while.

Opposite: A cattle roundup at Douglas Lake Ranch in the 1970s. *Item I–17010 courtesy Royal BC Museum*

Above: Joe Gardner, general manager of Douglas Lake from 1979 to 2019.

Above right: Clydesdale draft horses were still used at the Norfolk Ranch in the sixties. Joe Gardner, a teenager at the time, was taught to drive a team of the large horses.

"I did not have a real cowboy hat," recalls Joe, "so I shaped my Boy Scout hat to be a cowboy hat, and that resulted in my nickname being 'Scout.'"

He worked on the fencing crew for a few days, then was advised that he would be going to the Norfolk Ranch with foreman Fred Reimer the next day.

"Fred had a grip that could crush your hand, and with all the parts, groceries, mail, and other things already in the cab, I was relegated to the back box of the pickup for the drive to Norfolk. First stop was the Whites' place at Salmon Lake, to deliver their mail and supplies. I remember the mosquitoes being so thick I could hardly breathe.

"There were sixteen of us working at Norfolk. At the time, we were still using some Clydesdale teams. Norfolk was the only part of the ranch still using Clydesdales. We had a Clyde stud in the barn area. I was taught to drive a team and did so with dump rakes, horse mowers in swampy areas, and a sloop, hauling bales to the stack-yard.

"By the time I came back for a second summer, I was part of the family, with Fred, his wife Joan, and their three kids, John, Christine and Bill. I was trusted to take Fred's company truck down to the Home Ranch to pick up supplies. I remember one trip where I was told to go to the shop and get a pail of 90-weight oil for the horse-mower gear boxes. On arrival at the Home Ranch shop, famous at the time for having a cranky mechanic, I was asked, 'What do you want?' After telling the mechanic, he asked if I had a requisition—which, of course, I did not. No requisition, no oil, I was told. I was forced to go back to Norfolk without the oil, which drew a very angry response."

After three summers working at the ranch, Gardner entered UBC. While attending university, he worked evenings, weekends and holidays at a gas station near his home on the west side of Vancouver. The owner of the station asked him to start a tow truck service, which paid him more than he earned pumping gas but resulted in him being on call, carrying a pager day and night, and sometimes required him to skip classes to tow a vehicle.

Gardner finished his bachelor's and master's degrees in Agriculture, then started work for the BC Artificial Insemination Centre in Surrey, BC. His next career move was to Kamloops, where he became the district agrologist for the federal Department of Indian Affairs. When that office was closed, Gardner interviewed for a number of public service jobs in Canada and overseas, but he didn't really want to leave the Kamloops area. He eventually accepted a position as development manager for Calgary-based Abacus Construction, who were building townhouses and shopping centres in the BC Interior. While at Abacus, he gained valuable business knowledge and experience that would become important in his role as general manager of Douglas Lake.

IN 1979, WHEN Neil Woolliams announced his departure as manager of Douglas Lake, the vacancy attracted a large number of applicants. This was, after all, the premier job for any ambitious individual who aspired to lead the largest ranch in Canada into the future.

Joe Gardner emerged as a prime candidate. He had impressive credentials, including a master's degree in Agriculture, with research and a thesis on ruminant nutrition; work experience that covered all aspects of the job; work history on the ranch; as well as connections with people close to the owner and the ranch who verified that he was the one to fill the position.

"I was a late applicant for the job but was short-listed and called for an interview with Chunky and Carol Woodward," says Gardner. "The interview went well."

"The dialogue between Joe and Chunky was easy," remembers Carol Woodward. "My husband was a very private person who found it easy to relate to some people but not others. There was a comfortableness between him and Joe, and how they communicated. I could see them working easily together." Chunky valued his wife's observations and no doubt took her comments into consideration in his decision to hire Gardner.

"I was offered the job but had to keep it secret for a couple of months,"

Gardner remembers. "When Woodward was ready to make the announcement, he invited me out to the ranch and brought the management team together at the 'Big House' [Woodward's home on the ranch]. When he introduced me as the new manager, the first person to come over and shake my hand was cow boss Mike Ferguson, who congratulated me and said he'd support me. Some others were quite obviously not so enthusiastic. I learned later that some of the others in the room also had applied for the job and were disappointed they didn't get it."

On April 23, 1979, Joe Gardner began his career as general manager of Douglas Lake. In March 1980, he married Sandra Jean (nicknamed Sam) Stein. "We got married in our home at the ranch, with all the relatives," Joe recalls. "In the sunny morning, we had a game of mixed doubles tennis with Chunky and Carol Woodward, but by the time the relatives arrived, it was snowing, and by the time the marriage ceremony started, two inches of snow had fallen."

THE FLYING GM

When Gardner came to Douglas Lake as the general manager, he had a commercial fixed-wing pilot's licence and a Cessna 185 floatplane. Before joining the ranch, he had done some part-time flying for Central British Columbia Air Services and actually done some cattle-search flying for Douglas Lake.

Being able to fly and having his own plane allowed Gardner to fly to Vancouver to meet with Chunky Woodward, find missing cattle, and keep tabs on Douglas Lake's far-flung properties.

"Chunky didn't like my flying," says Gardner. "He thought flying in general was dangerous. However, he had a corporate jet and flew back and forth all the time. One time, he threatened to ground me, and we had one of those discussions where I said that would be a bad idea with a potential result neither of us would like. He backed off and, in fact, purchased the plane and relieved me of all the expenses."

When Chunky's friend the late Hugh Magee came up to the ranch, he came in his Cessna 185 amphibian floatplane. Magee and Gardner quickly became friends and flew together in various aircraft. When they eventually sold their Cessna 185s, Magee decided to get his helicopter licence and purchased a Bell Jet Ranger. Gardner decided to upgrade as well. He obtained his commercial rotary-wing licence and purchased a rebuilt Hiller UH-12E helicopter, though he sold it a few years later. For many years, Magee and Gardner shared the Bell Jet Ranger. Magee left it in the ranch hangar in the fall, winter and early spring, and took it to Vancouver the rest of the year.

Joe taking off in his Cessna 185 floatplane.

Below: Joe flying low in the helicopter he and Hugh Magee shared to move stray horses from the ranch's private grasslands.

Previous pages: *Yuki Sageishi*

HUGH MAGEE, FRIEND OF DOUGLAS LAKE

Hugh Magee, former CEO of Great West Steel Industries and GWIL Industries Inc., director of numerous for-profit and not-for-profit organizations, and mentor to many young business executives, was a long-time friend of Douglas Lake and Joe Gardner. His association with the ranch started when Chunky Woodward was the owner and endured through all subsequent owners. For years, Magee and Gardner talked regularly to each other on the telephone.

"My kids and grandkids don't know anything else but being a part of the ranch," Magee said. "They love the cattle, love agriculture. We all want to be here. My interest is the people. Some of my best friends work at Douglas Lake."

Magee visited his home on the ranch regularly between 1983 and 2019, when his failing health prevented further visits. In the early years, he flew his plane or helicopter, not only for transportation to and from the ranch, but also to help with whatever could be done on the ranch from the air.

The log cabin where Magee spent his time at Douglas Lake was built by Dick and Myrtle Reese in 1982. The Reeses were trappers who dealt with the ranch's problem wildlife—beavers, for example.

John West, Kip Woodward and Hugh Magee. *Courtesy Sherri Magee*

THE JOE GARDNER ERA 1979–2019

"While we liked the game on the ranch, and the water storage capabilities created by the beavers was useful, sometimes their numbers and dam-building got ahead of us, and we had to lower the population," says Gardner. "We wanted them to store water for cattle watering, but we didn't want them to plug our irrigation ditches and dams. They occasionally got a little carried away and dropped a tree on the front lawn."

The original cabin was built at the Sawmill Lake (also known as Rush Lake) cow camp and was upgraded in 1979 when an International Boy Scout Jamboree was held there. The Jamboree organizers installed power to the area and hauled in sand to create a swimming beach. "That was the year I became manager, and my predecessor, Neil, had agreed to the Jamboree," says Gardner. "Having been a Scout myself, I was surprised they had to have power. Even now, I meet people who say they were at Douglas Lake for that Boy Scout Jamboree."

Dick and his son built a beautiful new log cabin overlooking the lake, but the Reeses lived there for only a year or two before Dick died of cancer. His widow, Myrtle, stayed for about a year, but she didn't want to live there on her own, so she moved. After that, the ranch leased the cabin to Magee.

Cabins at Sawmill Lake, Hugh and Sherri Magee's Douglas Lake home. *Courtesy Sherri Magee*

MURPHY, MCKENZIE AND MILLIKEN

Prior to being hired as the general manager of Douglas Lake, Gardner had worked part-time for the company hired to conduct aerial searches for cattle on the ranch, so he was familiar with the "lay of the land" from the air. "During my first spring/summer riding with cowboy foreman Stan Murphy and his crew on the west end, and cowboy foreman Jerry McKenzie and his crew on the east end, I learned the lay of the land on the ground, as well," says Gardner. "That was very important."

Jerry McKenzie started working as a cowboy on the ranch in the early 1970s. He and his wife, Linda, lived at the Home Ranch with their three children—daughters Laura and Dale, and son Ryan. Jerry left the ranch in 1997 to become the cow boss at Nicola Ranch, but years later Ryan became a building contractor and worked on many projects at Douglas Lake, including renovations at Quilchena Hotel.

"Jerry reminded me that one day, when we were fencing, we had to use axes for clearing debris because the horse had bucked off the chainsaw and broken it," says Gardner. "According to Jerry, I went back to the Home Ranch that night and came back in the morning with a new chainsaw."

IT WAS DURING the summer of 1979, while Gardner was cowboying on Stan Murphy's crew, that he met Terry Milliken, who had been on Stan's crew since 1974 and, before that, had cowboyed at the Gang Ranch. "I worked at Douglas Lake for twenty years, and that was the best job I ever had," says Terry Milliken.

When the movie *The Grey Fox* was being filmed at Douglas Lake (see "In the Movies," page 153), and the production crew was looking for extras, Terry was one of the cowboys to volunteer.

QUARTER HORSE OPERATIONS

By the time Gardner was appointed general manager, Woodward had owned the ranch for twenty years and was ready to shake things up a bit. One of the first changes was to hand control of the Quarter Horse operations to Gardner. For years, Woodward had handled the registration of the horses from his Vancouver office, but moving all of that to the ranch made the most sense.

During the 1980s, Woodward hosted cutting competitions at the ranch, with upwards of 200 riders arriving in recreational vehicles hauling horse trailers. Woodward was a skilled competitor, sometimes riding Peppy San, the world champion cutting horse owned by Douglas Lake at the time. In

Quarter Horses freely graze in the large pastures of Douglas Lake Ranch. *Yuki Sageishi*

1986, the legendary Texas cutting horse expert Matlock Rose, who was Peppy San's trainer, came to the Douglas Lake competition.

The Quarter Horse operations had outstanding debts that needed to be collected, so Woodward took advantage of Gardner's tough nature and made him collector. "That made me unpopular," Gardner admits.

"My time signing horse registrations for all our foals each year, for thirty-plus years, made my signature and me well known by the various departments of the American Quarter Horse Association and staff at their headquarters in Amarillo, Texas. I served on some committees for a short time and always enjoyed visiting the headquarters and staff."

The Quarter Horse operations were an important part of the ranch during Woodward's ownership years and beyond, culminating in the 2004 American Quarter Horse Association's Best Remuda Award for overall quality of the ranch's horse herd. (A remuda is the herd of horses from which cowboys choose the horses to be used for the day.) The ranch also won a couple of world champion saddles and numerous trophies for their Quarter Horses.

The Quarter Horse shows and cutting competitions are no longer held

THE JOE GARDNER ERA 1979–2019

Characterization of "Rock Creek" Tom Dynneson, by artist John Schnurrenberger.
Courtesy John Schnurrenberger

at Douglas Lake, but the bloodlines and foundation work Woodward did remain to this day in the 400 or so ranch Quarter Horses. These horses play a crucial role in the ranch's successful cattle operations. All cattle work is done on horseback, and cowboys are on horseback every day.

For many years, the University of Saskatchewan's Western College of Veterinary Medicine sent an instructor and students to assist in calving with first-calf heifers. "Groups of veterinary students and their instructor, notably Dr. Eugene Janzen, came to assist at our English Bridge facility," says Gardner. "This led to huge improvements in calving results for us and a lot of practical hands-on experience for the vet students. Douglas Lake provided housing, meals, horses, and some help with transportation. This was also a valuable learning opportunity for our cowboys."

The arrangement ended when genetics, nutrition, pelvic measurement of replacement calves, and bull selection improved the percentage of live calf births, and when the cost of hosting the students—many of whom were going into small-animal practice—at the ranch escalated.

THE SAME YEAR that Gardner started as general manager, Tom Dynneson, better known as "Rock Creek," because that's where he was from, started working as a cowboy at Dry Farm.

Dynneson also worked at the Quarter Horse barn where his daughter, Leanne, remembers, "He had to deal with Sanctuary, the stallion, who on occasion would charge you if you were on a saddle horse."

Dynneson met his wife, Sharon McCrae, when she was working as a cook at the Quilchena Hotel. When she and Tom got together, she started working at the cookhouse at the Home Ranch and then as a camp cook at Chapperon.

GARDNER WAS NOT always the easiest boss to work for. Shortly after he started as general manager, Gardner fired a young employee for not being honest. "One day, I was driving out of the Home Ranch driveway and coming toward me, driving a company vehicle, was Kathy Gabara, who was working at the Quarter Horse barn at the time," he remembers. "I stopped her to say hi and have a short visit. I asked her if she had a driver's licence, and she said yes. When I returned to the office, I checked and found out she was fifteen, so I went looking for her and told her she was fired for lying to me, because no one could get a driver's licence until age sixteen. Kathy ended up marrying one of our young cowboys, Tom Ednoste (aka 'Forty'), and eventually we rehired her. To this day, we're still friends and she still jokes about being fired."

Kate (Kathy Gabara) Hendrickson's first visit to Douglas Lake was when she was thirteen years old. Leah and Harold Dinsdale (farm crew) invited her to stay with them through the school spring break. "It was calving time at English Bridge," Kate remembers. "It was a very exciting time in my life, to be invited to the largest ranch in Canada. Meeting cowboys seemed like a fantasy from a John Wayne movie, and I was right in the middle of it all. A weekend chore position at the Quarter Horse barn became available and I applied. I was fifteen at the time. Tony Burke hired me on the spot. Who knew mucking out barn stalls would capture my heart and soul?"

EFFICIENCY IMPROVEMENTS

Owner Woodward wanted to improve ranch efficiency and accomplishing that was up to Gardner. At the time, crews and equipment were based at Norfolk, Minnie Lake, Chapperon, Harry's Crossing, and Home Ranch, and two additional crews moved between smaller cow camps. "Foremen from the different ranches were always arguing and wouldn't help each other," Gardner remembers. "Each claimed ownership of his particular part of the ranch and was possessive rather than cooperative. I tried improving communications, but small changes weren't having much success."

Gardner suggested to Woodward that all parts of the ranch be consolidated under one foreman. With the owner's agreement, Gardner consolidated the farming operations and centralized management and equipment at Home Ranch, as much as possible. As a result, some of the former foremen left, some retired, and Arnold Nielsen was appointed farm boss.

Equipment that didn't work well was replaced with more efficient machines that could do more in less time with fewer people. "We had some twenty-seven tractors, largely Internationals, and it seemed six or seven of them were always at the shop being repaired or waiting for parts," Gardner recalls. "Finally, I couldn't stand it any longer and replaced those twenty-seven with nine new John Deere tractors. We lined up all the old tractors in the Home Meadow, added other equipment, and held a big auction. That went very well and allowed us to reduce operating costs and do a better job."

Gardner played an important role not only in the local, provincial and national cattle industry, but also in the international industry. In 1981, the Canadian Cattlemen's Association organized a national tour for a delegation from China, including a stop at Douglas Lake.

Joe and Sam Gardner, centre, with a delegation from China at English Bridge.

Rescue on Thin Ice

"One bitterly cold minus-20-degree morning in December 1982, I drove to Kamloops for a Kamloops Exhibition Association board meeting," recalls Gardner. "On arrival, I got a message to call my executive assistant, Candice Roulston, as soon as possible. On doing so, I learned that cowboy John Young had gone up to Hatheume the previous day to get some cattle that were stuck up there and that he had not returned. I headed back to the ranch immediately. I had my Cessna 185 warmed up with heaters, and Orval Roulston, a cowboy volunteer

who knew the country, was waiting to fly with me. Orval knew that country better than anyone else, from his years of experience as the cowboy foreman covering that area. Douglas Lake was socked in, but I knew from my drive from Kamloops that it was clear blue above the layer over the Douglas Lake airstrip.

"We took off, punched up through the clouds, and in a few moments we were scouting the area around Peterson and Pennask lakes, where we knew John was going. While we could see tracks where he would have gone to bring the cattle out, they ended, and we could not find him or his horse in the area. We knew if he got into trouble, he would have headed to our Chapman's cabin on the shore of Hatheume Lake, where he could find shelter and warmth.

"After circling several times, we could see no smoke coming from the cabin, so we went down low over the lake and looked under the tree cover at the cabin. John's horse was standing by the cabin, saddle on, but there was no sign of John. After a few more passes (a 185, power on, makes a horrible racket), we saw

Chapman's cabin at Hatheume Lake, where John Young was rescued. In later years, Gardner, with the help of brother-in-law Norm Stein, built a wood-fired sauna at Chapman's cabin, which was used by ranch people snowmobiling in that area.

John stagger out on the wharf. He was not waving on the next pass, but he was there. On the following pass, he fell face-first in the snowbank on the partially frozen lake. With John face-down in the snow at minus-20-something, we knew we did not have time to get back to the ranch to launch on snowmobiles, so we decided to land on the partially frozen lake. We were on wheels, and fortunately the ice held. We had to get out to turn the plane around, as it could not turn on the ice. We then started to taxi back to where John was.

"John was alive but unable to speak. He was very weak and cold. Our options were to take him in the cabin, get a fire going and warm him up, or fly him out. We chose to fly him out. We literally had to stuff him lengthwise in the plane as we couldn't bend his legs. Orval decided to ride the horse back to the ranch.

"I took off, the ice held, and I flew directly to Kamloops, which took fifteen or twenty minutes. Of course, I had the heat on full in the plane, but it was still very cold. We were met at Kamloops by an ambulance, and the attendants immediately started working on John. Once they left for Royal Inland Hospital, I flew back to the ranch and landed under the layer. By the time I got back to my office, we already had word that John was recovering.

"John made a full recovery. We learned that his horse had slipped and fallen on the trail, and John had hit his head, suffered a concussion and kept passing out, so when awake he would light a fire and then pass out again. When daylight came, he decided his horse would take him back to Chapman's, where he knew he could get shelter and we would find him.

After being rescued from a remote cabin in freezing winter weather, John Young decided to cowboy in Hawaii.

"Interestingly, after that experience, John decided to cowboy for a while at the Parker Ranch in Hawaii.

"As the story got out, Orval and I got some recognition for what we thought was just doing our jobs. First, Mayor Mike Latta, a friend of Chunky's, gave us a City of Kamloops plaque, and then we received a Workers' Compensation Bravery Award. A short time later, we received a plaque and medal from the Carnegie Hero Fund Commission, which also came with some cash. In 1986, Orval and I were both awarded the Medal of Bravery. This involved a trip to Ottawa for a ceremony at Rideau Hall and an award presentation by Governor General Jeanne Sauvé. Actually, wearing the medal or using the MB behind my name didn't happen for another twenty years."

EMPLOYEES—THE FABRIC OF THE RANCH COMMUNITY

Many of the people who Gardner hired remained at Douglas Lake for decades.

WENDELL STOLTZFUS was working at a nearby ranch and set his sights on working for Douglas Lake. "There was talk of a new manager, Joe Gardner, up at Douglas Lake," remembers Stoltzfus. "March 1, 1983, I hired on at Douglas Lake on the cowboy crew, fulfilling my boyhood dream." Stoltzfus was promoted to cowboy foreman on the west side a few years after he joined the ranch.

Stoltzfus met and married his wife, Leann, at Douglas Lake. "Our wedding day at the Portland Ranch was a hot July day when we broke the heat record," he remembers. "All the little evening church services and potluck suppers were special times for me and my family [daughters Madison and Reesa were born and raised on the ranch] during our twenty-four-year stay at Douglas Lake. When the helicopter arrived, I really enjoyed the rides, looking for cattle in the high country in that little glass bubble. Especially memorable was the trip to Amarillo, Texas, to receive the 2004 AQHA Best

Cowboy John Young, who was rescued from near-death in 1982.

"Got both heels," says Gardner.

Tim and Christine O'Byrne at Douglas Lake in the mid-1980s.
Courtesy the O'Byrnes

Remuda Award. There were fun team-roping days, too. Joe and I actually caught and won some money one time. I remember specifically one time at the new arena at the Quarter Horse barn, Chumley bucked Joe off. We had many fun times in that arena."

IN 1984, WHEN Arnold Nielsen retired, Stewart Murray, who had already been working at the ranch for seven years, was appointed farm boss, a position he still holds (as of 2023).

Murray was born and raised in Victoria, BC, but spent most of his summers at his grandparents' ranch in Idaho. Murray studied diesel mechanics at Camosun College, in Victoria, but found few jobs available upon graduation. When he saw a posting for a job at Douglas Lake, he applied and was hired, in 1977, to work on the irrigation crew and to drive tractors. He became the lead hand at Home Ranch, then farm boss. "I have no real job description," Murray says. "I do a bit of everything, taking care of the mechanics, truck drivers, farm and irrigation crews. I'm responsible for feed production, equipment, some building, infrastructure such as roads, waterlines, sewers."

Murray met his wife, Sheila, at a harvest party, and they were married in 1980. Sheila's father had been foreman at the Norfolk Ranch. When needed, Sheila worked in the cookhouse. They raised three daughters on the ranch. "There were lots of kids here back in those days," Murray remembers. "The place grows on you. The biggest thing here is that there's always something new, something new every day."

Murray was involved in more than farming operations at the ranch. "In the early 1980s, we purchased a used cement truck and built a new set of feed grounds on the 'Island' that had a capacity to feed some 2,500 backgrounding calves through the winter," says Gardner. "Stew Murray was the main guy on the project, and Dale Arnell worked on the design and construction, and then further improvements to the English Bridge feed grounds. We also built a feed mill and a large concrete trench silage pit."

The "Island" is a large field at English Bridge that is surrounded by Chapperon Lake to the east, Nicola River to the south, and Chapperon Creek to the north. "The feed ground looks like a feedlot, which it could be, but the difference is that it's only used for backgrounding calves through the winter rather than feeding cattle year-round or fattening them," Gardner explains. "We designed the feed ground to hold 5,000 head, but only built pens for 2,500, along with the processing facilities and scale. The facilities paid for themselves because we bought calves in the fall that we could background, then either sold them as grass yearlings in the spring or put them on leased grass and sold them in the fall. One place we sent these cattle was to the 100

Mile Ranch, where for years we dealt with Peter and Marina Castonguay. Another major improvement was the construction of a new 60-by-100-foot steel shop with a large overhead door to accommodate even our large trucks and equipment. The shop had a hoist, parts room, tire room, and a wash rack outside, and had a used oil burner for heat to back up the propane heaters."

TIM O'BYRNE JOINED Douglas Lake in 1984. "Highly respected camp boss Orval Roulston hired me on at Dry Farm," recalls O'Byrne, who is now editor of Nevada-based *Working Ranch Magazine*, which has a circulation of more than 60,000 beef cattle producers in North America. "The crew were pretty much top hands, all of 'em. What I learned from these men impressed me for the rest of my life. The real gem of my time there was discovering the almost legendary management of the extremely fragile and valuable deeded native bunchgrass grazing holdings the ranch has in abundance. Venerable cow boss Mike Ferguson took his responsibility very seriously, one handed down to him from his predecessor, Ol' Joe Coutlee, and for decades he solemnly protected those grasslands like they were his own.

"Under cow boss Stan Jacobs's mentorship, I eventually took on my first foreman job at Chapperon division. What I lacked in ability, I made up in a commitment to do my very best for the two and a half years that I, my wife Christine, and son Mark were there before moving on to the Gang Ranch under Larry Ramstad.

"Standout memories include the absolute Cowboy Heaven I enjoyed with the boys, including Trevor Thibeault, Kelvin Grabowsky, Terry Milliken, Wendell Stoltzfus, Miles Kingdon, and even a bit with Jake Coutlee; of getting ready for Panorama Sale and then riding proudly on the big day and half the night with twenty pots (cattle liners) lined up on the road in front of the chutes waiting to load that evening; pulling a calving shift with 1,400 heifers at English Bridge; Joe in his helicopter, fluttering above my outstretched carcass after I got bucked off a three-year-old; and noisy dancing sandhill cranes courting on the edge of Chapperon Lake during spring thaw.

"What I recall most about Douglas Lake was the pride everyone took in just being a part of it all—from the owner, C.N. Woodward, on down, across Joe's desk, to the Quarter Horse barn, the farm crew, fence crew, shop and irrigation crews, the cooks, to the office, general store, and the cowboy crews scattered out over 550,000 acres. The horses were mighty fine, the equipment excellent, but my impression of that Douglas Lake pride in meticulously keeping the outfit in tiptop shape sticks out most in my mind. I consider myself blessed to have been allowed a backstage pass into the inner workings of this amazing cornerstone of North American ranch life."

Opposite: Feeding cattle at English Bridge.

FOR FIVE YEARS, from 1986 to 1991, Duncan Barnett was Gardner's assistant manager. Barnett had worked as a summer student while attending UBC in Agriculture and joined full-time as assistant manager after he graduated. "Duncan was very thorough at any project on which he worked," says Gardner. "He was deliberate and detailed. He worked on our banking arrangements and our complicated insurance needs. He worked on various springs, including Wilson Springs, documenting flow and other data. We later named an unnamed spring behind Dairy Lake as Duncan Springs." Gardner also recalls Barnett's detailed analysis of truck makes and models when the ranch needed a new feed truck.

"I worked on field sales of fertilizer, salt and fencing, and we had memorable sales trips," Barnett remembers. "I may have driven Stewart and the farm crew crazy with my analysis of the cost of feed production."

Jane Barnett, Duncan's wife, clearly remembers the day she was driving back from work in Merritt along Nicola Lake bluffs when a huge boulder fell and hit her car. She was pregnant with their second daughter at the time. Fortunately, while the incident was stressful, Jane was uninjured. Duncan, Jane and their family left the ranch in 1991 and bought a ranch outside of Williams Lake.

IN 1989, WHEN long-time cow boss Mike Ferguson announced his retirement, Gardner selected Stan Jacobs as Mike's successor. Jacobs had been working at Douglas Lake since 1987, initially as foreman at Portland cow camp, and continued as cow boss until 2019.

Jacobs grew up around cows and horses near De Winton, Alberta, and rode bareback in rodeos for a few years. He worked at the Alberta Livestock Transplant Centre, then as a farrier, and in 1980 moved to BC. After working at some of the large ranches in the province, he set his sights on Douglas Lake. "I was looking for a ranch that was well established and not going to break up," said Jacobs. "I talked to Joe, and told him when there was an opening, I'd be interested." On November 7, 1987, Jacobs started his long history with Douglas Lake.

Much changed during the years that Jacobs rode the ranges on the ranch, but the constant always was, and still is, that the cowboys do all their work on horseback. "The complexity is when calving and branding changes with the seasons, even though it's all predictable," Jacobs said. "We're still reliant on the environment. As one of the fellas said, if we weren't working here, we would not get to see any of this—the most gorgeous sunrises, the best and worst cattle drives—we wouldn't be fortunate enough to see any of it."

"Stan's legacy at Douglas Lake is the cow herd and the genetics he selected," says Gardner.

Left: Long-time cow boss Mike Ferguson (left), who retired in 1989, with west end foreman Stan Murphy.

Below: Stan Jacobs, ranch cowboy since 1987, cow boss from 1989 to 2019.

IN 1989, TREVOR Thibeault joined Douglas Lake as a cowboy and he, his wife, Bernice, and son Clay moved to Chapperon cow camp. "Like most young cowboy families, we were very poor but excited to start working for the biggest cow outfit in Canada," Thibeault remembers. "After being there a year, my foreman quit, and I was offered the job, and that was the start of a mission for me and my family, which soon included our younger son, Casey. I was extremely focused on making the Chapperon end of the ranch productive. Joe wasn't always patient with me, but Bernice and I always appreciated his straightforward approach. You always knew where you stood with Joe. 'That didn't work well' is a quote that I learned from Joe. There was a lot of advice in that comment that I still use to this day when one of my ideas doesn't pan out.

Trevor Thibeault, foreman at Chapperon in the 1990s.

"As time progressed and we got things organized, Joe gave us Hatheume cow camp to look after along with Chapperon. Bernice became our camp cook, which she excelled at."

Bernice tells a story about the time she and sons Clay and Casey went with a number of other ranch families to find the perfect Christmas tree: "We went to Big Meadow. We had just started out when Clay noticed a hole under a burnt-out old stump and said, 'It looks like something is living in there.' Casey shot by me and jumped into the hole and was smiling and waving to everyone when one of the other ladies yelled that there was a bear in there. We all headed for the truck, but Casey and the other lady fell coming down the hill. I stopped to help her up and turned to see the big brown bear running just as hard as we were running—in the opposite direction."

The Thibeaults left the ranch, and Trevor now has a contracting company in Cache Creek. "Douglas Lake is one of the greatest ranches in the world," says Trevor. "Our time there was a very unique experience that molded us into the people we are today."

Cathy Lewis with a jar of Rocky Mountain oysters, aka bull calf testicles.

"Chunky Woodward always insisted that we have help at home so that we would be available to spend time with him and his wife when they were at the ranch, and to entertain guests," says Gardner. "Sam and I saw an ad in the Merritt paper that Cathy Lewis was looking for a job. Cathy was from St. Kitts, in the West Indies, and was visiting her sister, who was working in Princeton.

"Cathy was shy and timid when she first started working with us, respectfully addressing us as Mr. and Mrs. Gardner, but as time went on, she became more comfortable with us and jokingly would tell me, 'Go to hell, Joe.'

"Since she couldn't drive, I decided to give her driving lessons. First thing she did was drive into a ditch. She didn't realize she had to turn the wheel. It was a harrowing experience, but she did eventually get her driver's licence.

"After about three years, Cathy moved to Kamloops, brought her two sons from St. Kitts to Canada, became a Canadian citizen and bought a house. She still lives in Kamloops and is in regular touch with us."

THE JOE GARDNER ERA 1979–2019

Previous pages: Horned Hereford Bull. *Yuki Sageishi*

Gardner was invited to speak to the British Cattle Breeders in 1998 at King's College, University of Cambridge, England.

THROUGH GENETICS AND NUTRITION

With Woodward's agreement, Gardner changed feeding and breeding programs, which had dramatic results in selling prices and weight gains for the ranch's cattle. Yearling steers that sold at Panorama sales increased from 600 pounds to 1,000 pounds. "We did that through genetics and nutrition," says Gardner. "Mike Ferguson and then Stan bought superior bulls. That was the most important factor in improving the herd."

Over the years, Gardner has been active in various cattle associations, including as a director of the Nicola Stock Breeders Association; director and president of the BC Livestock Producers Co-operative Association; president of the BC Association of Cattle Feeders; first president of Ownership Identification Inc., BC's brand registration and inspection program; and trustee of the Canadian Cattlemen's Foundation (now Canadian Cattle Foundation). To keep up to date on the industry, he also attended events such as the National Cattlemen's Beef Association's annual convention for US beef producers. "I figure when you're dealing with the largest herd of beef cows in Canada, you should pay attention," says Gardner.

VIP VISITORS DURING CHUNKY WOODWARD'S OWNERSHIP

Chunky Woodward's high profile in the business world and recognition as owner of Canada's largest cattle operation led to Douglas Lake hosting royalty, in addition to Prince Philip and various VIPs. Some of the memorable visitors include:

The Duke of Westminster
"Chunky called me and said he had some special guests coming to the ranch on the company jet," Gardner remembers. "Since he couldn't come up, he wanted Sam and me to host them. I had no idea that the special guests were British royalty—the Duke and Duchess of Westminster [Major General Gerald Cavendish Grosvenor, 6th Duke of Westminster].

"The Duchess [Natalia] was tall and wore tight purple jeans and penny loafers. The Duke had long hair combed to one side, with a bobby pin to keep his hair out of his eyes. This was not what we expected of such regal guests. I was surprised to discover that Gerald Grosvenor had, as a young man, worked at the ranch one winter. Of course he wanted to show his wife where he had worked and stayed, so off to Chapperon we went. In the old, abandoned bunkhouse, he found where he had carved his initials on the wall. When we tore down that bunkhouse, I saved that section, framed it, and shipped it to him.

"We had a good time with them and later learned that Grosvenor was one of the richest men in the world!" Grosvenor was chair of Grosvenor Group, one of the world's largest privately owned property businesses, started by the Grosvenor family in 1677.

Lieutenant Governor Henry Bell-Irving
In May 1982, Chunky and Carol Woodward invited BC's Lieutenant Governor, Brigadier-General Henry "Budge" Bell-Irving, to visit the ranch for a week-long ride. "It was quite a production, with staff moving the camp each day and setting it up at the next location," Gardner recalls. "The large group included Budge and his wife, Nancy; two aides-de-camp, Dick Vokel and David Harris; Kamloops Mayor Mike Latta and his wife, Emily; Mike Ferguson; Chunky and Carol; Sam and me; and others.

"Budge and his wife started at the Big House. Budge rode on an English saddle, which was most comfortable for him. Some meals were back at the Big House, where the non-riders stayed.

"One dinner party at the Big House that Sam and I particularly remember was with Budge, Chunky, Mike, and Austin Mitchell, who were telling war stories. At the time, Mitchell was at the ranch fishing with Sam's father, Gordon, and staying at our home. Turned out that Austin had been Budge's driver during the War. (Budge earned his Distinguished Service Order and Bar during WWII, leading the Seaforth Highlanders of Canada in Italy and Holland, and was promoted to brigadier.) One weekend during the War, when Budge was being picked up and flown back to London, Austin asked if he could use the Daimler armoured vehicle they had liberated to go to the nearest village for Friday night. Budge agreed. When Austin and others came out of the party, the Daimler was gone. It had been stolen. When Budge's return flight landed Sunday afternoon, Austin picked him up in an army jeep. He was very concerned about what was going to happen when he told Budge about the Daimler. Budge's reaction was 'Oh well, we will have to liberate another one.'

"The last day, we met with Chief Herbie Manuel and the Upper Nicola Band, and that evening we were hosted at Corbett Lake Lodge [20 km southeast of Merritt], where proprietor Peter McVey threw a magnificent dinner for local ranchers and businesspeople."

Governor General Edward Schreyer

In the mid-1980s, Governor General Edward Schreyer and his wife, Lilly, were guests at the ranch. Schreyer was the 22nd Governor General of Canada, from 1978 until 1984.

"His Excellency did not relax much and was rather stiff, despite Chunky and Carol's best efforts and touring," Gardner remembers. "Carol was preparing a special dinner and, when ready, asked the governor general's aide de camp to advise him that dinner would soon be ready. The aide did not wish to disturb him. Carol, of course, prevailed and then softened him up with a couple of her special Bloody Marys. I was then able to get him interested in some agricultural discussions. I remember he was interested in soil science and, in particular, peat."

Rick Hansen, Man in Motion

Wheelchair athlete Rick Hansen and his wife-to-be, Amanda, visited the ranch in May 1987, just prior to the scheduled May 22nd completion of his Man in Motion World Tour in Vancouver. He took a one-day break and was greeted by the Douglas Lake school students when he arrived at the airstrip.

"Chunky took him fishing," Gardner remembers. "Rick is an avid outdoorsman and returned to the ranch many times after that."

Above: Amanda and Rick Hansen with Chunky Woodward.

Left: Sam Gardner between Malcolm Forbes and Chunky Woodward. "Surrounded by money," quips Joe Gardner.

Malcolm Forbes

"In the summer of 1986, we had some interesting visitors at the ranch," recalls Gardner. "An entourage of Harley-Davidson riders led by Malcolm Forbes [publisher of Forbes magazine] arrived for lunch. At the time, Forbes was rumoured to be dating Elizabeth Taylor. Travelling with the group was our Kamloops MLA at the time, Claude Richmond, who was also the Minister of Tourism."

Scotland Yard Commissioner Sir David McNee

"In the early 1980s, we had another special guest, Sir David McNee and his wife, Isabel," remembers Gardner. "McNee was, at the time, commissioner of Greater London's Metropolitan Police Service, also known as Scotland Yard, the largest police force in Britain. They were being toured through the Interior of BC by the RCMP, who were treating them like royalty, Red Serge and all, guards around the Big House, and police escorts on the road.

"The respect shown McNee and his wife was interesting, but when Sam and I hosted them for a couple of hours at the Big House, we had a great visit. Sam had been worried about protocol and serving tea with the silver service, but when she asked Isabel what she would like, the couple asked if we had any Scotch, which of course we did.

"McNee told us the story of a recent event where he was in charge during the siege of the Iranian Embassy in London, which his force handled until the first hostage was shot. At that time, McNee turned control over to the British Army, which deployed the Special Air Service to storm the building and resolve the situation. I remembered seeing news reports of armoured forces swinging in through the large windows from ropes on the roof."

Chunky Woodward Dies

"On a typical weekend in the spring, Chunky and Carol would fly up to the ranch," says Gardner. "Sam and I would join them for drinks and dinner, and maybe some cards, usually hearts, which was difficult with Carol, as she could count cards, and she soon knew what everyone was doing and would sometimes announce it.

"One particular weekend in April 1990, the weather was good, so we played doubles tennis, Sam with Chunky versus Carol and me. Chunky was making some serious gets, and we all expressed concern that he was going a little too hard for someone who was going in for a scheduled surgery on Monday. He advised us that he was fine and that his doctor had told him it was okay.

"Chunky's surgery went well, but later that week, while he was still in the hospital, he suffered heart arrhythmia. He died on April 27, 1990. What a shock. He was only sixty-six years old and otherwise in good health.

"About a year earlier, Chunky had announced to me that he had made me the vice-president of the cattle company. When I asked if that meant more pay or different duties, he said no. I asked again what the new title meant, and he said it meant that I was vice-president. With his passing, that of course meant that I was the only officer of the company while the estate was being settled."

CHUNKY WOODWARD'S DEATH was sudden and unexpected. "I was at the ranch the day Dad died," remembers son Kip Woodward. "I drove back to Vancouver so fast I got a speeding ticket." Kip told the police officer that his father had just died and that he was rushing to the hospital, but the officer said he'd heard that story before and gave Kip a ticket anyway.

Woodward's death came at a time when Woodward's department stores

At the Celebration of Life for Chunky Woodward, his extended family and employees were entertained by western music icon Ian Tyson.
Courtesy Woodward family

were facing the end of a century of retail success in Western Canada. John Woodward, a Marketing graduate of the British Columbia Institute of Technology (BCIT), had been executive vice-president and chief operating officer of Woodward's, while brother Kip, a Business and Economics graduate of the University of Western Ontario, had been vice-president of Merchandising.

The Woodward sons started their ownership responsibilities in the midst of a landmark battle between Douglas Lake and the provincial government over the use of land granted under a grazing licence. "When the Ministry of Transportation started planning the connector (Highway 97C, also known as the Okanagan Connector or Coquihalla Connector), we were involved at the onset, as it was clear the highway would pass through some of our land, no matter which route was chosen," Gardner remembers. "The first routes proposed went through just south of the Home Ranch. Our pushback eventually relocated the route to its present alignment. To deal with the expropriation, we hired a lawyer who had extensive experience and produced good results for other ranchers during the first two phases of the Coquihalla Highway. We were soon before the Expropriation Compensation Board and making what could be best described as extremely slow progress. Meanwhile the road was under construction. Compared to the claims of ranchers before us, ours was very large. The provincial government of the day passed new legislation referred to as the 'Douglas Lake Amendment,' wiping out the largest part of our claim. We felt this was outrageous, and we protested the grand opening with cowboys, cattle and signs. The RCMP tasked with security actually came to us to ensure that we were not going to block the road. We learned from that experience that precedent and the law of the land can be changed."

"Overnight, we put up huge protest billboards that would be seen by politicians and media when the highway opened," remembers Kip, who found the process a learning experience. "I learned how to navigate around government."

As much of a distraction as the Coquihalla Connector dispute created, Gardner and the next-generation Woodward family had a ranch to run. "We focused on cattle," Kip says. "That was our number one business. We decided we needed to grow the cattle business. Forestry was our number two business. We wound up the Quarter Horse cutting and showing businesses, sold some of the top cutting horses but kept the main horse herd and breeding program. We focused on the major assets and profits."

Douglas Lake protest sign at the opening of the Coquihalla Connector.

A SHORT-LIVED VENTURE IN DEER FARMING

Chunky and some business buddies from Vancouver had also started a fallow deer farm on about 300 acres at the Home Ranch. "Venison was going to be the next big thing in restaurants," says John. However, during the couple of years after the Woodward sons and daughters inherited the ranch, the deer market slumped. "At one time we had imported fallow deer from England," John adds. "While deer antlers were very big in the Chinese medicine trade, we never did that because it was too tough on the animals. We just sold deer for breeding as well as for meat."

The fallow deer farm at Douglas Lake as it appeared in late 1980s and early 1990s.

The deer farm was managed by Dale Arnell, who started working at Douglas Lake in 1989. "It was a big learning curve for all of us," says Gardner. "Dale soon became a good deer master."

"I treated them like any other livestock," says Arnell, who had no previous experience with deer. "They were considered exotic game, so we were required to contain them behind high fences."

When the Woodward brothers recognized that the costs of keeping the deer exceeded the value of sales, they recommended to the partners that they close down the operation. Gardner found a Texas buyer for the entire herd.

MULTI-TALENTED DALE ARNELL

Arnell left to work at a large feedlot in Alberta but returned to Douglas Lake as cattle buyer. "Dale would attend the fall cattle auctions in the Interior of BC and try to buy calves that we could background," says Gardner. "If he bought calves, he would advise us, and we would send one of our cattle liners and meanwhile get everything ready, including the feed and lighting, so that when the truck arrived late at night, the cattle could go straight onto feed and water, then be processed and received the next morning."

Arnell also worked on special projects. He had a drafting program that enabled him to go out to the fields with his GPS [Global Positioning System] to create a rough image of the pastures. "I walked miles and spent weeks doing that for all pastures," says Arnell. "Then I spent about a week with a GIS [Geographic Information System] company in Vancouver that translated

Magy and Dale Arnell.
Alana Miller, Highpoint Design

all my information into GIS data. They told me to get proper equipment and do it right, and Joe said to go for it, so we bought the hardware and software to do it. I'm self-taught."

Arnell is self-taught in a number of different skills. "Dale tracked feed consumption in our backgrounding lots and also started what became our major invasive-weed control program," says Gardner. "He has also been in charge of our radio-frequency identification ear-tag program since its inception." In the early 2000s, traceable ear tags became mandatory, and as of spring 2019 Arnell had registered and kept track of more than 150,000 tags.

"Dale's work allowed us to have a Verified Beef Production Plus herd, which provides proof to consumers and retailers that the beef cattle operations adhere to the highest standards for food safety, animal care and environmental stewardship, and to reap the benefits of that program," Gardner adds. "As well, he tackled other construction and design projects, such as mapping the legal corners of all our land and registering all our wells."

"I'll do anything I'm asked to do," Arnell says. "I can fit in anywhere." Arnell created the maps in this book. He and his wife, Magy, live at Chapperon.

CARL WOODMAN, EQUIPMENT OPERATOR EXTRAORDINAIRE

Carl Woodman started working at Douglas Lake as an equipment operator in 1990. "There was no equipment he could not operate and, furthermore, at an expert level," says Gardner. "As we got to know him better, he became even more valuable, as we could give him a project like building a bridge across a river. We gave him the materials and manpower he needed, then simply stood back and watched him get the job done.

"Most bridges on the ranch were built by Carl, and all construction projects were capably managed by him. He could supervise a crew or get a job done himself. He left his mark on many projects, including dams, roads and feed lanes. Several balanced rock piles that can be seen about the ranch were built by Carl with an excavator.

"A few years after joining us, Carl met Rebecca, a young Swiss chef who was working for us at Stoney Lake Lodge. Carl's son Sam went to the Douglas Lake school and worked for us as a cowboy for many years. Carl and Rebecca left in 2013."

NEW BUSINESSES

With so many lakes on the ranch property and an ever-increasing number of trespassers poaching the fish, the Woodwards and Gardner decided to turn recreation into a business. They started breeding fish to stock the lakes, then developed facilities for fishers and charged fees for day-users and overnighters.

Gardner had also started a wholesale business selling products that every ranch used, including fertilizer, fencing and even equipment. "We sold the provincial government a lot of the fencing for the Coquihalla Highway," says Kip Woodward. "Everything we used, we could buy in greater quantity for less than smaller ranches could, so we wholesaled these things." The ranch put its equipment into the Highways Department's hired-equipment list when not in use on the ranch.

Gardner and the Woodwards partnered with Sanders Contracting on projects such as building dams and bridges, earthmoving, and other construction jobs. One memorable project for the partners was the Chilko Lake Spawning Channel in the Cariboo. "This was our most remote project and a very intriguing job," says Jerry Sanders. "The helicopter flight to tour the site was an adventure for me, especially when we ended up in the middle of a gigantic thunderstorm on the way home. I know what popcorn feels

Neil Blackwell became the ranch's purchasing agent and general store manager in August 1994. Blackwell sources and purchases everything for all the operations, including seed for the farm; feed for the livestock; salt for the cattle; supplies for the cookhouses, camps and store; stationery for the office; and everything else.

"My mother-in-law saw an ad in the newspaper for this job," says Blackwell. "The ad said running a general store and purchasing. I sent in a resume, and Joe called me for an interview. I was surprised there were several people in that interview, including Joe, Stew, Stan and others. They all asked me questions. I arrived in a suit and tie, and at first, I didn't know who Joe was, so I addressed all my answers to the guy in a shirt and tie. Turns out he was the accountant. I finally figured out Joe was the one in cut-offs and flip-flops.

"Every day is different. This is not your typical nine-to-five job. I just get started on something and someone gives me two more things to do. There's so much variety. The number one thing, though, is lifestyle. The general store closes for a half-hour for lunch, so I even get to go home for lunch." While he still goes to Kamloops or Merritt to pick up supplies once a week, nowadays larger purchases and groceries are delivered.

Blackwell, his wife, Gail, and three children have lived on the ranch for over twenty-five years.

like. We had a few hair-raising adventures. Who could forget the sunken excavators here and there, notably the one at Douglas Lake's dock?"

The partners decided to form a highway maintenance company, and for a few years in the early 1990s WhiteLine Road Maintenance was the biggest in the area, eventually securing contracts in the Kamloops and 100 Mile districts. Gardner, Kip and John Woodward, Jerry and Ron Sanders, and Dave Cunliffe were partners in the venture, and Cunliffe was the first general manager of WhiteLine. "We'd all known Cunliffe from his days as district manager of Highways in Merritt," says Gardner. "While he was in that position, he made many improvements on and about the ranch and the Douglas Lake Road.

"After about four years, we no longer enjoyed the highway maintenance business and made some money selling out. Cunliffe continued to work as a consulting engineer for Douglas Lake."

As smart, energetic, enthusiastic young businessmen, Kip and John Woodward could not have had a more compatible manager for their ranch.

"Joe had great entrepreneurial flair," says Kip. "We discussed ideas with him, and he would run with them. Joe's a smart guy, and he was getting bored. He'd been doing the cattle thing for years. We wanted to diversify, and we had the person to do it with. I like to think that, for Joe, it was an exciting time. He was firing on all twelve cylinders. We were always scheming."

In December 1993, Jim McGill joined Douglas Lake as financial controller, a position he held until his retirement in January 2010. "During my tenure, there were three owners: the Woodward family, Bernie Ebbers, and Stan Kroenke," says McGill. "My time at Douglas Lake was the most interesting I've had in my working career as an accountant. After all, how many people ever get the opportunity to be a part of a truly unique operation with the historical significance of the Douglas Lake Ranch?"

CLOSING THE WOODWARD CHAPTER

In mid-May 1995, neighbouring Upper Nicola Band members blocked the access road to the ranch in a dispute over traditional fishing and hunting rights. Leaders of other Bands and supporters from around the province joined the protest. After two weeks of negotiations, an agreement was reached, and the blockade was removed.

During the negotiations, the Woodwards learned from the federal and provincial governments that Indigenous rights were yet to be determined. Rights were evolving. "John and I found out that neither the federal nor the provincial government could explicitly support private property," says Kip. "Ranches depend on government land for grazing, and those grazing rights come up for renewal every ten years. The big picture was changing for us."

As years went on, more and more factors led the Woodward siblings to conclude that ranching was not the best business for the family. "We built the business as much as we could without debt," says Kip. "Joe and the senior managers were a good team. We grew the cattle business, the tourist business, and the sales business. It was running just fine. But our family had tax issues, there were macro issues—land and property rights issues—and between the four siblings, we had no one in the next generation who wanted to take over from us."

"The four siblings agreed it was time to sell," adds John Woodward. "The dot-com industry was booming, and we hoped to sell the ranch as a trophy asset."

"In the context of the day, people were making money on technology, we thought the existing business would look good to those buyers," says Kip. "But the asset produced low return. We knew it was going to be difficult to sell."

The Bernard Ebbers/ WorldCom Connection

In 1994, when the Woodward family decided to sell Douglas Lake Cattle Company, they recognized that finding a buyer would be a challenge and enlisted the assistance of the corporate finance professionals of KPMG, the accounting firm that the family had worked with for decades. Al Kanji, KPMG senior partner in charge of mergers and acquisitions at the time, headed up the team that searched the globe to find a buyer.

"Kanji was the gatekeeper and negotiator, ensuring that only those who were judged capable of purchasing the ranch could receive detailed information about the operations," says Gardner.

"The ranch is not something an individual would buy based on it being a commercial venture," says Kanji. "This is a unique asset, a trophy asset. Finding comparisons would be difficult. The buyer would have to love it—similar to buying a painting.

"There was a limited ultra-high net-worth horizon of potential buyers, and our firm had access to those unique people who might be interested. We made private, discreet enquiries—one-on-one contact—through our firm's network of offices around the world. There were only twenty to twenty-five potential buyers in the UK, Europe, the US, Canada, and some parts of the Far East. This was the 1990s, and we didn't have the same access we do now to some of the Asian countries.

"To put it into perspective, Douglas Lake is bigger than some countries. That was appealing to some people. Another appealing aspect was that the value of the ranch was made up of specific components—cattle, agriculture, forestry, recreation—in addition to the value of the privately owned property.

"There wasn't immediate interest. It was a long-term process, and the Woodwards were patient and realistic. They were keen that the legacy of the ranch be kept. We relied heavily on Joe for his knowledge of the ranch. After all, it was, in a sense, his ranch."

There was a potential US buyer interested in the property, but after almost two years, there were no meaningful offers. "One day, I got a call from an Alberta rancher advising that he had a friend who might be interested," says Gardner. "That friend turned out to be Bernard Ebbers of WorldCom fame."

Telecommunications titan Bernard Ebbers, co-founder and chief executive officer of Mississippi-based WorldCom, contacted Kanji.

"Ebbers was not on our list," says Kanji. "Ebbers had read in an industry publication that the ranch was for sale. At the time, we had another party interested, but Ebbers was the most proactive. When he said he was going to do something, he acted on it. He was very easy-going, very easy to talk to, but I could see a demanding streak in him, and he was quick and decisive."

"Ebbers came for a tour with his girlfriend, Kristie," Gardner remembers. "Ebbers called her 'ETB,' meaning Ebbers-to-be. They liked what they saw and, in short order, made an offer." The Woodward family accepted the offer, and the ranch was sold.

"From the time Ebbers first came to us until he closed the deal took only three or four months," says Kanji.

Ebbers completed the purchase of the ranch in July 1998.

The change in ownership from the Woodward family to Ebbers was a time of uncertainty for Gardner, who had been general manager of the ranch for almost twenty years. "I was unsure as to where the sale would leave me, but Bernie asked me to stay on," says Gardner. "I could not have been treated better by the Woodwards. John and Kip and their sisters treated us very well."

One of Ebbers's first additions to the ranch was the church at the Home Ranch. The church is still a prominent building that serves multiple purposes. Visiting pastors have held services, and the facility has been used for weddings, funerals, and other religious events.

West Bay Construction, which had previously built facilities at Douglas Lake, built the church. "I met Cliff Oughtred and his wife, Margo, at our place at Silver Star," says Gardner. "Cliff was a well-known ski coach and had a construction business in Kelowna. West Bay rebuilt the Salmon Lake facilities, including Twigg's Place, Stoney Lake Lodge, the new office, two new houses at the Home Ranch, and a new home for Carol Woodward at Chapperon after Chunky's passing. Cliff's crew, known as the Blue Meanies, worked hard and fast and did good work."

The bell at the top of the church is a replica of the bell from a church in the Kootenays that fell on the head of the bellringer, killing her, in 1998.

Bernard Ebbers was born in Edmonton, Alberta, and grew up there, as well as in California and New Mexico. He earned a basketball scholarship to Mississippi College, and upon graduation with a degree in Physical Education, he started his career as a teacher and basketball coach. For a while, he owned and operated hotels in Mississippi, then in the 1980s became involved in the telecommunications business. He started small but grew his organization into the giant WorldCom. Ebbers earned significant fame and fortune, and by the late 1990s was estimated to be worth more than a billion dollars. Much of his personal holdings, including the Douglas Lake Cattle Company, was backed by WorldCom stock holdings.

At the turn of the millennium, Ebbers and WorldCom began to unravel. A class-action civil suit and then a criminal investigation led to charges of conspiracy, securities fraud, and filing false reports. This was the largest accounting scandal in US history at the time and resulted in billions of dollars in losses to investors, the bankruptcy of WorldCom, and the loss of almost 20,000 jobs. In 2005, Ebbers was convicted of charges against him and, at age sixty-four, sentenced to twenty-five years in a minimum-security prison in Oakdale, Louisiana. In late 2019, he was granted compassionate release, and on February 2, 2020, Ebbers died.

During a visit to Douglas Lake, Ebbers proposed to Kristie on the bridge over the Nicola River, on the road to the Horse Range.

THE BERNARD EBBERS/WORLDCOM CONNECTION

Douglas Lake church, completed in 2000.
Yuki Sageishi

Cliff Oughtred found the bell, borrowed it, and had a foundry in Penticton cast two replicas, one of which was sold to pay for the one he installed at the top of the Douglas Lake church.

"EBBERS WASN'T HERE as much as Chunky had been, but he and his wife, Kristie, flew up regularly on a corporate jet, often with her daughter Carli and family friends," Gardner remembers. "He was keen to get to know everyone who worked at the ranch, know their names and their families. They liked to get out and about on the ranch, attending any and all functions."

Ebbers had used his shares of WorldCom as security to buy the ranch, and as part of the 2002 corporate bankruptcy, the ranch was sold.

"When WorldCom began to unravel, Bernie met with me many times to explain in detail what was going on," says Gardner. "He did not believe he was guilty of anything. Be that as it may, he was convicted, and we were thrown into a court-ordered sale process that created worldwide press coverage."

With the media exposure that both Ebbers and WorldCom attracted, there was no shortage of potential buyers. "A press release was issued worldwide announcing that Douglas Lake was going to be sold," says Gardner. "I had phone calls from people around the world saying they wanted to buy the ranch.

"As of July 2002, KPMG, Douglas Lake's accountants since before I became general manager, were no longer allowed to continue as our accountants. A fellow from AlixPartners in Chicago was sent to take control of the ranch for the courts. He assumed that I was Bernie's man and could do things that would harm the value of the ranch. He flew back and forth each week and asked a million questions. At some point, the AlixPartners representative determined that I was actually looking after the ranch and most likely was the one who could help them sell it. After that, he kept in touch but did not continue flying back and forth each week.

"As the sale process was launched, the US real estate people and AlixPartners arrived for a tour. I remember well touring them. They knew absolutely nothing about ranching. I took them by the calving barn, where we had a crew calving our first-calf heifers. It turned out my daughter, Taylor, was on shift and on horseback, and was just bringing in a heifer that needed help with calving. We stood back and watched as Taylor caught the heifer in a stanchion, put on a plastic glove, shoved her arm in the heifer, attached some obstetrical chains, and with some heaving on both her and the heifer's parts, pulled out the calf, which she then weighed. Shortly thereafter came the afterbirth, which was attacked by her dog, and that was followed by the real estate guys puking all over the place. It was hard not to laugh. Taylor had a live, healthy calf and heifer, and I had a very sick bunch of guests.

"Wealth was the only common factor among the people we toured through the ranch. Some wanted to get rid of the cattle and run buffalo, others were more interested in the ecosystem and animals, while still others wanted to know if we could extend the runway and build a hangar for their 737.

"One of the groups did not like the suggested Vancouver drop-off point for bids and decided to have their business manager hand-deliver the bid personally, fifteen minutes before closing. He landed by helicopter on the front lawn of my house and delivered the package to me. Interesting times."

Stan Kroenke, who had looked at the ranch with interest around the time when the Woodward family accepted Ebbers's offer, was still interested and had kept in touch with Gardner. "There was a bidding process, but nobody knew what Ebbers had paid for the ranch," Gardner says. "There was speculation, but no one knew. As the highest bidder, Kroenke signed a deal in late 2003 and completed the purchase in early 2004."

The Stan Kroenke Era

When sports and entertainment billionaire Stan Kroenke purchased Douglas Lake in 2004, the property consisted of some 164,000 deeded acres and about 500,000 acres of Crown grazing licences. The purchase also included significant timber resources, cattle, horses, and machinery, as well as several locations of Douglas Lake Equipment.

The announcement of Kroenke's purchase cast a celebrity aura over Douglas Lake, which soon cooled because of his intensely private nature.

"The sale happened when the industry was in the middle of a BSE [Bovine Spongiform Encephalopathy] wreck, with cattle values swinging wildly and affecting the value of working capital in the company," Gardner remembers. "Closing was delayed, but did occur, and once again we had a new owner.

"While I was aware of Stan's wealth and his interest in sports teams and commercial real estate, I found him very easy to be with and deal with. When he is at the ranch, he likes to tour about and see even the far-flung operations and distant corners. He has made major investments in the ranch, with large capital improvements such as pivot irrigation systems and equipment upgrades.

"Whenever I became aware of other opportunities, he was supportive, and we were able to purchase some other fine ranchlands. I worked for Stan longer than any previous owner. He purchased other large ranches in the United States during the same time period and is now a substantial owner of ranchlands and very well established in the cattle industry in Canada and the United States. I believe he sees great value in large grassland ranches."

STARTING IN 2008, Douglas Lake Cattle Company began a series of acquisitions to expand their land and operations. Most of the acquisitions

Douglas Lake has owned Douglas Lake Equipment since 1999, when Bernie Ebbers purchased the failing New Holland farm equipment dealership in Kamloops. Douglas Lake Equipment grew rapidly, selling farm and construction equipment and trailers, and expanded to Vernon, Surrey, Williams Lake and Quesnel. In 2004, one of Kroenke's first expansions was to Douglas Lake Equipment, purchasing Kenver Equipment in Dawson Creek, BC, and Grande Prairie, Alberta, when owner Ken Haverland was ready to retire.

In 2008, with a downturn in the economy, sales of trailers and some short lines of equipment were discontinued, and operations in Vernon and Williams Lake were closed.

Gardner oversaw the operations of Douglas Lake Equipment until 2009, when Gary Frelick was hired as the new president. While the new and used heavy equipment sales and rental company is still owned by Douglas Lake Cattle Company, the operations are separate. Douglas Lake Equipment now has outlets in Surrey, Kamloops, Quesnel and Dawson Creek in BC, and Grande Prairie in Alberta.

have been the result of succession challenges experienced by the families who previously owned the ranches. As the younger generation chose careers other than ranching and moved away from the family home and business, their parents had no choice but to sell.

"The ranches we acquired were not on the market," says Gardner. "As general manager of Douglas Lake, people interested in selling or buying ranches talked to me. During those conversations, I became aware of opportunities.

"In buying the ranches we did, we increased our cow herd and reduced operating costs. Mr. Kroenke wanted to expand, and we agreed on the criteria. We wanted large, deeded acres of grassland. Douglas Lake is the crown jewel, with thousands of acres of natural grasslands—this is one of the largest remaining habitats of bluebunch wheatgrass.

"Grassland ranches are the lowest-cost producers because the cattle can eat the grass and turn it into beef. Available grass reduces the need for feed. Most of our pastures are naturally well-watered, and over the years we have added dugouts and other water features where needed. Many ranches have less deeded grasslands, so they have to feed their cattle more days of the year. All the ranches that we have acquired have extensive grasslands; that was the attraction."

Most ranches in the province have access to Crown grazing ranges. "These Crown acres can generally be accessed only between June 15 and October 15,"

Above: *Brent Gill*

Gardner explains, "and that leaves five to six months when the cattle have to be fed. If there are no deeded grasslands, that's a lot of feed. We look for big chunks of grassland. The owner is not interested in development on these lands. He's interested in cows, because the ranches make money with cows.

"What Douglas Lake has now is perfect. That's not to say that something won't come along that might pique the interest of the owner, but it would have to be a considerable opportunity, and it would have to make sense. At the moment, there is nothing on the radar. For now, Douglas Lake is concentrating on doing things efficiently."

ACQUIRING GRASSLANDS

The ranches acquired by Douglas Lake more than doubled its land base and the number of cattle. Each ranch played an important role in British Columbia's ranching industry, and over the years each has had interesting connections with Douglas Lake.

Alkali Lake Ranch

In April 2008, Douglas Lake purchased the historic Alkali Lake Ranch, BC's oldest cattle ranch, located 50 kilometres southwest of Williams Lake, in the Chilcotin. At the time, this ranch added 37,000 deeded acres and 125,000 acres of grazing permit to Douglas Lake's land base. Ten years later, Alkali Lake's

livestock consisted of roughly 1,300 cows, 350 replacement heifers, and 55 horses, with a year-round crew of six cowboys and six farmers. Employee housing for families and a bunkhouse and cookhouse for singles are provided.

Doug and Marie Mervyn had owned Alkali Lake Ranch since 1977. One of the key reasons they purchased the ranch was the grassland ranges, and for that same reason Gardner initiated the conversation that led to Douglas Lake buying Alkali. With four children and a growing number of grandchildren, the Mervyns chose to sell the ranch to ensure that the property would be kept intact and would not be subdivided. "We wanted the grassland to be continued as grassland, and Douglas Lake does a good job of grasslands," says Mervyn.

"Alkali Lake Ranch was a major expansion for us," says Gardner. "To be able to make the deal on 37,000 acres of historic ranch was a feather in Mr. Kroenke's cap, a major feat!" (For more information on Alkali Lake Ranch, see Appendix A, page 172.)

Gardner at Alkali Lake Ranch, overlooking the Fraser River.

Circle S Ranch

In 2012, Douglas Lake acquired the Circle S/James Cattle Company at Dog Creek. Circle S is adjacent to Alkali Lake Ranch and added almost 20,000 deeded acres, 2,300 leased acres, and access to 203,000 acres of Crown land.

"Since Circle S was our immediate southern neighbour, I visited Lyle

James on occasion," says Gardner. "At one point, James had the ranch listed for sale. While Circle S was no longer listed, my friend and realtor Barry Cline visited James and recognized that the ranch was in decline, as was James's health and ability to keep it going. Mary, Lyle's wife, was his main help. The acquisition of Circle S was a natural progression for us." (For more information on Circle S Ranch, see Appendix B, page 174.)

Quilchena Ranch

In 2013, a year after purchasing Circle S Ranch, the unexpected happened. Douglas Lake Cattle Company acquired the Quilchena Ranch, which included 28,000 deeded acres, grazing rights on 61,000 acres of Crown land, the historic Quilchena Hotel and Restaurant, a general store, RV park and golf course, all adjacent to Douglas Lake. "I never thought Quilchena would ever come to us," Gardner says of the acquisition. "Guy Rose and I didn't exactly see eye to eye, but he had some health concerns [Rose died in 2019], and I had told him if he ever decided to sell, I wanted him to call me first. About four or five months later, his accountant called me.

"To negotiate the sale of the ranch, Guy appointed Steve Tidball, a director of Quilchena Ranch and a member of the high-profile George Tidball family, who started Thunderbird Show Park and owned Keg Restaurants. Since we were both well enough known to cause suspicions if we were to be seen together, our meetings were clandestine, in hayfields and all sorts of secret places. Between us, we got the deal done.

Curt Martindale started working at Douglas Lake on the cowboy crew in 2002 as a teenager from Alberta. He continued for four years, then left to work at the Quilchena Cattle Company. When Quilchena became a part of Douglas Lake, Curt was back working for Douglas Lake again.

"Curt worked his way up the crew and in 2016 became the west end/ Portland cowboy foreman," says Gardner. "He also found time to concentrate on a local girl, Erica Huber, who became his wife. They live with their two children at Quilchena, on her parents Eugene and Sherri Huber's place.

"My favourite story about Erica is when she was working at the Quilchena Hotel. Stan Kroenke and others, including me, went to the pub for dinner one night. Stan asked Erica what the catch of the day was. She went to check, and when she came back from the kitchen, she said, 'It got away.' We all crack up every time the story is retold."

"Quilchena Ranch was another natural acquisition for us, with their land either just across the fence from us or completely surrounded by us. They had family members, managers, and an accountant, all of which we already had, so this substantially reduced operating costs."

When Douglas Lake bought Quilchena Ranch, they were not only in the ranching and forest management businesses, they were also in the recreation business—in an even bigger way than they'd previously been.

"Quilchena Ranch was another beautiful grassland ranch with a succession problem that came to our attention as a huge opportunity," says Gardner. (For more information on Quilchena Ranch, see Appendix C, page 177.)

Riske Creek Ranching

In 2015, Douglas Lake expanded again, this time buying Riske Creek Ranching Ltd.'s Cotton and Deer Park ranches, across the Fraser River from Alkali and Circle S ranches. The Riske Creek ranches now include 30,000 acres of deeded land and 157,000 acres of Crown grazing land. Some 950 cows are maintained on the ranches, and 700 acres are under cultivation.

"Grant Huffman, one of the owners of Riske Creek Ranching, called to ask if we might be interested in buying Riske Creek," says Gardner. "I had known Grant since our years at the University of British Columbia, in Animal Science, where we both did our master's degrees. We got the deal done, but it took some time on due diligence because there were some abnormalities in the survey of District Lot 44, which is one of the largest pieces of private property in BC, with more than 9,000 acres.

"The extensive grassland, with its scenic breaks and vistas of the Fraser and Chilcotin rivers, is a sight to see. They had pivot irrigation out of both the Fraser River and Riske Creek. Just off Farwell Canyon Road, they had the Company cabin site with a stack-yard, where they would stage hay for feeding in the winter and spring. The large spring, with water pouring out of the bank in an otherwise very arid landscape, made it a perfect site for a feeding ground and for calving a few cows. The old, unusable log buildings were burned down by a Forestry back-burn fire.

"Further down the road, the provincial Wildlife Branch has a pull-out and kiosk explaining the formation of the Junction Sheep Park in the 1970s. Many people go to this attraction and take the goat trail to visit the park. I've seen many sheep on our northern ranches. Further down the road is a very popular pull-out on our deeded land, with some signage where people can take a brisk, moderately tough hike out to the sand dunes overlooking the Chilcotin River." (For more information on the Cotton and Deer Park ranches, see Appendix D, page 179.)

Following pages: *Yuki Sageishi*

JP, Bobbi, Duke and Will Parkes. *Courtesy Bobbi Parkes*

In 2017, Bobbi Parkes joined Douglas Lake as Gardner's executive assistant. "Bobbi and her husband, John (JP), and kids lived at the Nicola Ranch, where JP was the manager," says Gardner. "When a house became available at the Home Ranch, Bobbi and JP and their family moved to the ranch and, instead of Bobbi commuting, JP commuted to Nicola Ranch. At one time, JP worked on the cowboy crew at Douglas Lake, too.

"Bobbi is a gem, with great social, computer and business skills. Her kids, Duke and Will, stopped by the office whenever the school bus dropped them off. She and Dana Gill made a great team in the office. One year, at the annual Christmas party they organized, Bobbi and Dana dragged me onto the dance floor to the tune 'Working for the Man.'"

THE EXPANDED DOUGLAS LAKE OPERATIONS

With more recreation facilities and the new northern operations, Gardner was ready to hire an assistant manager. It was around this time that Phil Braig's email arrived.

"I wanted to work towards a management position on a ranch in BC," says Braig, who had spent ten years during his high school and university years working at various jobs that had given him experience in all aspects of ranching. The addition of a business degree, he knew, would also help him to secure the job he wanted. "Joe's email address wasn't on the Douglas Lake website, so after looking at the ones that were, I figured his would follow the same format. I didn't hear back for a couple of weeks, then Joe called and said he'd call next time he was in Vancouver. He called in October, and the following spring, Brittney [Parks] and I went up for a visit. Joe toured us around. He said he was looking for someone to train to replace him and asked if I would be interested in doing that."

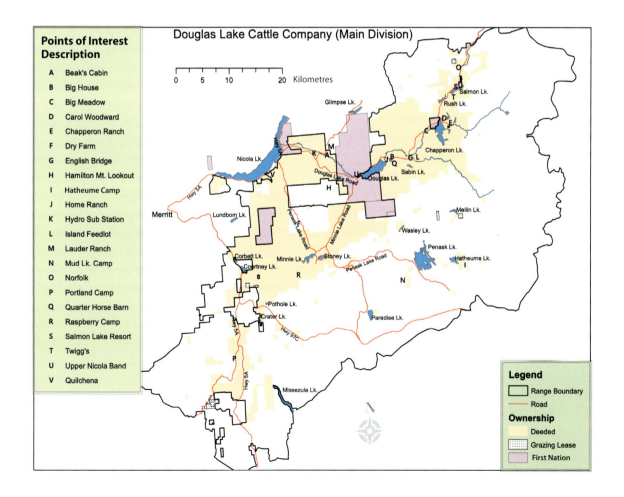

Douglas Lake's southern (main) operations. *Map by Dale Arnell*

THE STAN KROENKE ERA

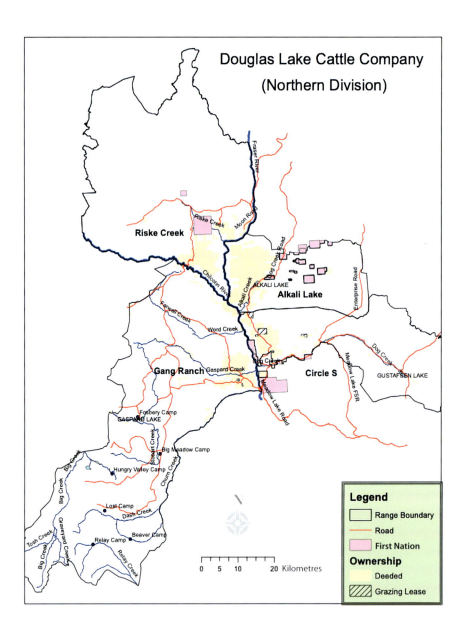

Douglas Lake's northern operations.
Map by Dale Arnell

For a city kid raised in Vancouver, Phil Braig managed to build a diverse background in horses and cattle. At age twelve, he spent a summer in the Clinton area, helping to take people on trail rides and learning to care for horses at a guest ranch that was also a working cattle ranch. He was enthusiastic and willing to do anything and, as a result, people were eager to teach him what he needed to learn. The following summer, he returned, and the year after, he worked on another ranch, east of Williams Lake. There, he learned about haying and operating equipment. Each summer, he found new work experiences, including training horses and cowboying in BC, Alberta,

Phil Braig, general manager of Douglas Lake Ranch since 2019. *Courtesy Douglas Lake Ranch*

and Saskatchewan. By the time he earned his business degree at Simon Fraser University, he knew he wanted to work on a ranch.

But first, Braig accepted an internship at the Canadian embassy in Berlin, then got a commercial helicopter licence and travelled through BC and Alberta, gaining experience and flying hours. "I realized that wasn't the career I wanted long-term," says Braig, who also has a fixed-wing pilot's licence. "I wanted to get back into ranching and find a way to combine my business degree with my ranching experience." That's when he contacted Gardner at Douglas Lake.

In April 2012, Braig started by working alongside the cowboys on the ranch, while also handling some administration tasks with Gardner. At first, he lived on the ranch during the week and spent weekends in Vancouver, where Brittney still lived and worked. While Brittney was on maternity leave, they moved to Merritt, then rented a house in Quilchena for two years before moving to Home Ranch.

In July 2019, Braig was promoted to general manager following Gardner's retirement from day-to-day management of the ranch.

THE STAN KROENKE ERA

Cattle are the number one business of Douglas Lake Ranch. *Yuki Sageishi*

INTEGRATING THE OPERATIONS

Over time, operations in all Douglas Lake locations have been standardized. "For example," says Gardner, "our cattle medication protocol was the same for all operations. With larger cattle numbers, plans are easier to design and implement."

Finding qualified people who want to live and work at the ranches, especially the northern operations, is an ongoing challenge. Finding seasonal workers is an even bigger challenge. Since 2016, the ranch has been participating in the federal government's Seasonal Agricultural Worker Program, which allows employers to hire temporary foreign workers when Canadians and permanent residents are not able to fill available positions. The participating country's government recruits and selects workers, ensures workers have the necessary documents and qualifications, maintains a pool of qualified workers, and appoints representatives to assist the workers in Canada. Douglas Lake ranches hire workers from Mexico.

Starting in January, Douglas Lake advertises for seasonal workers and might get one or two qualified applicants, but not enough to fill the job vacancies. As a result, foreign workers are hired at all operations. They work from April until the end of October, during peak irrigation, until the end of haying. Once they are no longer needed, they go home to their families. Douglas Lake pays for their return air ticket and provides accommodation in the bunkhouses while they are working on the ranch.

Cattle—The Number One Business

Douglas Lake is Canada's largest working cattle ranch. Its sustainably managed natural grasslands and water resources and efficiently farmed fertile lands place Douglas Lake ranches among the lowest-cost cattle producers in the country. "If we manage the land properly, we will have sustainability forever," says Gardner. "As long as we do not over-graze, we have a sustainable low-cost way to raise beef."

Douglas Lake ranches have a total of about 20,000 cattle and can support about 11,500 mothers. The base herd consists of about 4,500 Hereford cows, 2,500 Black Baldie cows, and 450 bulls, producing about 6,800 calves a year. The ranch's cattle have a reputation for exceptional quality and uniformity.

Cattle being corralled in preparation for branding. *Yuki Sageishi*

"We spend time on genetics, and our cattle work well on the range and in the feedlot," said Stan Jacobs, cow boss from 1987 to 2019.

In a typical year, Douglas Lake aims for a 90 per cent pregnancy rate. "Our bull-to-cow ratio is about one bull for every twenty cows," explains Gardner. "For our first-calf heifers, we strive for about six per cent bulls. That keeps the conception rates where they need to be. The bulls go out to the heifers around May 8 and to the cows around June 8.

"We try to buy above-average bulls for below-average prices. Bull selection and buying well are important tasks for the cow boss."

Calves are born starting in late February. From about mid-June to mid-October, cattle are turned out onto Crown grazing land. All Douglas Lake ranches are located in a climate where ticks can thrive. In the spring, cattle are sprayed to prevent disease, paralysis, or even eventual death that can result from tick infestations.

The cowboys' work is endless. The cattle must be moved between ranges not only to ensure that they have enough to eat and drink, but also to ensure that grasslands are not overgrazed. They monitor the health of the cattle and constantly check that gates are closed and fences are in good repair. From late April until early June, they brand the calves and move the bulls into the cow and heifer ranges.

Training young horses is also the cowboys' responsibility. Having one or more good cow dogs can be advantageous, and most cowboys on the ranch own and train dogs to help in their day-to-day work.

Cattle on the move in winter at Douglas Lake.
Brent Gill

When Stan Jacobs and his wife moved to the ranch in 1987, their son Cameron was just seven months old and daughter Megan was two. Cameron and Megan went from Kindergarten to Grade 12 at the ranch school. As soon as he was old enough, Cameron started cowboying, and once he graduated high school, he worked on the ranch full-time. He followed his father's footsteps and took a farrier course in Oklahoma, but being a farrier wasn't what Cameron wanted to do. He moved to Alberta, where he worked on ranches and at feedlots, then in 2009 moved back to Douglas Lake to cowboy at Chapperon. Two years later, he was appointed foreman at Chapperon. Cameron and his wife, Trina, who was working in the yard at the Home Ranch when they met, had two children and lived on the ranch until 2019.

"A typical day for the cowboys starts at 5:00 in the morning, when we meet for breakfast. By 5:30, we're pulling out of the yard," explained Cameron Jacobs while he was still working at Chapperon. "In summer, when it's hotter and lighter, we meet at 3:00 for breakfast. The cows move better when it's cool. We usually try to be done by 2:00 or 3:00 in the afternoon. Our average day is about ten hours, sometimes twelve or fourteen. We try to have a few seven-hour days to compensate."

Being a cowboy is more than a job, it's a lifestyle. "I love the life," said Cameron, "just being outdoors all the time in the peace and quiet."

Left: Cameron Jacobs participates in the bucking horse event at a ranch rodeo.

Above: Cameron Jacobs, Chapperon foreman from 2011 to 2019.

In the early fall, the cowboys move the cattle down from the higher Crown grazing lands to the grassland owned or leased by the ranch. These extensive grasslands enable the ranch to graze the cattle later into the fall and earlier in the spring than most ranches, thereby having to winter-feed for a shorter period of time, which results in significant cost savings.

Sometimes cattle are lost during a move and can end up miles from where they're supposed to be. Aerial searches help to identify the location of stray cattle so that the cowboys can retrieve them. Fortunately, the Douglas Lake brand is well known, and if cattle end up in another ranch's herd, they are easily identified.

Once the cattle are back down on the lower ranges, the calves are weaned, starting in late October. Between 400 and 500 calves are weaned at a time, and the cows are pregnancy-checked. Those that are pregnant are turned out to graze, and those that aren't are sold at local auctions.

IMPACT ON RANCHERS WHEN CATTLE ARE SUSPECTED TO HAVE DISEASES

All calves are ear-tagged so that their age and herd of origin can be tracked. "Barcode tags were started in the late 1990s and radio-frequency tags started in the early 2000s," said Stan Jacobs. "With the outbreak of BSE, traceability became important. If one of our cows shows up anywhere in North America, it can be traced back to Douglas Lake."

Disease and even the suspicion of disease in a cattle herd can have serious impacts on a ranch. "In June 2009, the Canadian Food Inspection Agency (CFIA) notified us that routine testing at slaughter identified one of our cows as testing positive for bovine anaplasmosis, an infectious blood disease usually spread by biting insects," Gardner recalls. "We did not believe we had the disease and had not seen any typical signs in our cattle, yet the CFIA placed us under quarantine and soon some of our closest neighbours, too. We had to scramble because they would not allow us to sell any cattle that were not tested, and we had some 1,000 head of yearling steers at the Dry Farm and yearling heifers at 100 Mile.

"Stan Jacobs and I worked through this operational disaster. Other cattle tested positive on our ranch and neighbours' ranches. Things became more doubtful when a public meeting was held and the doctor in charge of the laboratory started talking about false positives.

"When a neighbour's cow tested positive at first then later tested negative, we knew something was wrong. Fast forward through this ordeal: it all blew up when we found out that the original testing had been incorrect and that our and others' cattle did not have anaplasmosis."

Gardner had tracked every call and task and created a spreadsheet of costs resulting from the false accusations. With the help of politicians and the Canadian Cattlemen's Association, the ranch was finally compensated.

"When we got a call from CFIA in 2018 saying that routine testing had found a neighbour's cow with bovine tuberculosis, and that we, along with others, were being quarantined, I started another spreadsheet," Gardner says. "In our case, the CFIA wanted to test every animal on our place that was older than one year of age, or some 6,000 head. This was a massive project again, with some reactors to the first test requiring a follow-up test that would be more exact.

"We did lose some cattle due to a requirement that some positives be slaughtered for post-mortem inspection. The good news was that, when all was said and done, no tuberculosis was found in our herd. We were compensated for the animals slaughtered and those hurt or destroyed while testing. We also received, as other ranchers did, compensation for the ordeal. No spreadsheet needed. All other herds, except the index herd, were also cleared.

"The bad news was that the neighbour's ranch where the original case was detected did have other positives, and their herd was totally depopulated. Yes, they received compensation, but how do you rebuild a large, beautiful herd of cattle? This was a Century ranch [so named by the Province of BC to honour farms, ranches, or agricultural organizations that have been active in BC for 100 years] managed by the fourth generation of its founding family, and now they had to start over. How horrible!"

CATTLE MARKETING

Douglas Lake calves are "preconditioned," meaning they have all their vaccinations, have been dewormed, and have been fed rations in feed bunks for a minimum of three weeks before being sold. Preconditioned calves are a selling feature. The cattle are also "one-iron," meaning Douglas Lake has bred them, owned them since birth, and looked after them until they're sold.

Most Douglas Lake cattle are sold to feedlots in southern Alberta, though some go elsewhere in Canada or the US. "Prior to my arrival in 1979, the ranch sold some of its beef in Woodward's Food Floors in BC and Alberta," says Gardner. "The advertising featured pictures of Mike Ferguson, cow boss at the time, and our cattle. I remember news reports of people mobbing the Food Floors for the beef. The only problem was a shortage of supply. We applied for and got a trademark for 'Douglas Lake Beef,' which we used on a few occasions, including at Safeway, once Woodward's Food Floors were sold to Safeway." Douglas Lake Beef is not presently sold outside of the ranch.

A small number of cattle are sent to a nearby abattoir for the ranch's own

Panorama Sale auction at Douglas Lake, 1992, with Chunky Woodward in the middle of the photo, wearing the white hat.

use and for sale in the Quilchena general store and the store at Home Ranch. Douglas Lake Beef is aged twenty-one days and labelled as BC Verified.

In 1993, when Blair Vold, of VJV Auctions in Ponoka, and Brant Hurlburt, of Fort Macleod Auctions, formed the Canadian Satellite Livestock Auction, Gardner sold Douglas Lake cattle in one of the first satellite sales. Prior to that, buyers and auctioneers came to the ranch. The cattle were sometimes scattered in fields miles away and had to be driven for days to reach English Bridge, where they were then gathered into pens of similar-sized animals for the auction. Weight losses during the process were significant.

"On auction day, the trucks were lined up," remembered Stan Jacobs. "After the sale, we started weighing, and then loading the trucks, sometimes through the night and into the next morning. It was more stressful for the cattle and cowboys."

"Hundreds of people would come, but only two or three were buyers," Gardner remembers. The free barbecue and excitement of a livestock auction drew the crowds.

Now the process is online. The cattle are gathered and sorted by sex, colour and size, and then videoed. On auction day, the videos are displayed on the Internet and on screens at the auction yard, where an auctioneer accepts bids from the ring or via telephone or Internet link, and the cattle are sold to the highest bidder. Once the cattle are sold, trucking is arranged, and on shipping day the cattle are weighed and inspected. The process is easier on the cattle. "Douglas Lake cattle sell well because we have large numbers of cattle that are uniform, and how they perform and gain is consistent and predictable," said Jacobs.

Cows that are not pregnant or otherwise suitable to keep are still taken to a local live cattle auction, such as BC Livestock or the former Valley Auction.

SEASONAL CYCLES

The seasonal cycle at Douglas Lake repeats each year. In the spring, the cowboys calve and brand, while the farmers prepare the fields and irrigation systems, and seed the fields.

In the summer, the cowboys turn the cattle out onto Crown land, and the farmers start to harvest hay and silage. In the fall, the cowboys gather the cattle, wean the calves, and pregnancy-check the cows, while the farmers complete the harvest and winterize the irrigation systems. In the winter, the cowboys move the cattle to feedlots, sort the calves for sale, and find any stragglers on Crown land.

While feeding continues through January and most of February, this is also the only time of the year when everyone can focus on maintenance and repairs. Calving starts in February, and then the cycle repeats.

Left: Mike Ferguson branding at Big Meadow in 1993. The annual Big Meadow branding was usually held on the third weekend of April. In addition to all the cowboy crews and their families, owners and some special guests attended, and the ranch provided a picnic lunch.

Following pages: Cattle being moved by cowboys in the winter.
Yuki Sageishi

CATTLE—THE NUMBER ONE BUSINESS

Precious Grasslands

One of the most important features of the Douglas Lake ranches—the one that drew early European settlers to these lands—is its indigenous grasses, particularly bluebunch wheatgrass and fescues. These types of vegetation are resilient to the cold winters and dry summers that characterize the climate at all Douglas Lake ranches. However, these grasslands are susceptible to damage by overgrazing and destruction by human activity. "People get frustrated with No Trespassing signs," says Gardner. "The general public does not realize the damage they are doing when they drive snowmobile, quad, motorcycle, or mountain bike on sensitive grasslands. It affects the quality, quantity and species of plants. It also opens up the soil for invasive species. For example, there are sections of the old wagon road to Kamloops where the wheel ruts have still not filled in."

Sound grazing programs have always been important to sustain the natural forage and prevent the infestation of other, less desirable grasses. About 50 per cent of grasslands are left ungrazed each year to keep them healthy, protect against invasive weeds, and to provide cover for the endangered wildlife species that thrive on the ranch, including the federally protected sharp-tailed grouse and provincially endangered American white pelican.

"Large acres of Douglas Lake land are locked-up and protected, with No Trespassing and No Hunting signs posted," says Gardner. "These areas are similar to wildlife parks, and various biologists study the wildlife there. We carefully manage access wherever we can.

Opposite:
Yuki Sageishi

"Shauna Jones, an ecosystems biologist with the Environmental Stewardship Division of the Ministry of Environment, approached us, wanting to survey our ranches and others for sharp-tailed grouse. We agreed, subject to some conditions, which included not making exact locations available to the public, in order to protect the birds and their habitat. I accompanied her and her workmate on a number of ground-proofing hikes to areas that had been discovered by helicopter observation. It was interesting to discover that the numbers of sharp-tailed grouse were larger than previously thought and that most of the sites in the area were on our private, protected and well-managed grasslands. While our northern ranches, namely Alkali and Riske Creek, have not yet been surveyed, I believe those areas will also have sharp-tailed grouse."

Opposite: A bald eagle perches on a branch.
Edb3_16 / Adobe Stock photo

DOUGLAS LAKE WILDLIFE AND HABITAT

Among the many wildlife species that can be found on Douglas Lake ranches are the following:

BIRDS

Grouse
Ruffed Grouse
Spruce Grouse
Sharp-tailed Grouse

Loons
Common Loon

Cranes, Rails & Allies
American Coot
Sandhill Crane

Grebes
Horned Grebe
Red-necked Grebe
Western Grebe
Eared Grebe
Pie-billed Grebe

Cormorants/Pelicans
American White Pelican

Herons
Great Blue Heron

Waterfowl
Tundra Swan
Canada Goose
Snow Goose
Green-winged Teal
Mallard
Blue-winged Teal
Cinnamon Teal
Northern Shoveler
American Wigeon
Ring-necked Duck
Canvasback
Redhead
Lesser Scaup
Common Goldeneye
Barrow's Goldeneye
Bufflehead
Hooded Merganser
Common Merganser
Ruddy Duck

Shorebirds
Spotted Sandpiper
Killdeer

Gulls & Terns
Caspian Tern
Black Tern

Pigeons & Doves
Mourning Dove

Owls
Barred Owl
Great Gray Owl
Great Horned Owl

Nighthawks & Allies
Common Nighthawk
Rough-legged Hawk

Swifts
Vaux's Swift

Kingfishers
Belted Kingfisher

Woodpeckers
- Red-naped Sapsucker
- Downy Woodpecker
- Hairy Woodpecker
- Northern Flicker
- Pileated Woodpecker
- Lewis's Woodpecker

Hummingbirds
- Rufous Hummingbird

Diurnal Birds of Prey
- Red-tailed Hawk
- Northern Harrier
- Bald Eagle
- Turkey Vulture
- Osprey
- Golden Eagle
- American Kestrel

Songbirds
- Olive-sided Flycatcher
- Willow Flycatcher
- Eastern Kingbird
- Violet-green Swallow
- Cliff Swallow
- Gray Jay
- Black-billed Magpie
- Common Raven
- Red-breasted Nuthatch
- Ruby-crowned Kinglet
- Swainson's Thrush
- American Robin
- Cedar Waxwing
- Grasshopper Sparrow
- Wilson's Warbler
- Yellow-headed Blackbird
- Common Yellowthroat
- Spotted Towhee
- Chipping Sparrow
- Savannah Sparrow
- White-crowned Sparrow
- Dark-eyed Junco
- Western Meadowlark
- Brown-headed Cowbird
- Purple Finch
- House Finch
- American Goldfinch
- American Avocet
- Gadwall
- Horned Lark
- Merlin
- Northern Pintail
- Shrike

Virginia Rail
Western Wood Pewee
Western Kingbird
Tree Swallow
Northern Rough-winged Swallow
Barn Swallow

Bank Swallow
American Crow
Black-capped Chickadee
Marsh Wren
Mountain Bluebird
Hermit Thrush
Bohemian Waxwing

European Starling
House Sparrow
Yellow Warbler
Yellow-rumped Warbler
Western Tanager
American Tree Swallow
Vesper Sparrow
Song Sparrow
Harris's Sparrow
Red-winged Blackbird
Brewer's Blackbird
Bullock's Oriole
Cassin's Finch
Pine Siskin
Evening Grosbeak
Common Snipe
Greater Yellowlegs
Long-billed Curlew
Mountain Chickadee
Semipalmated Plover
Sora
Wilson's Phalarope

Above: A great blue heron. *Gregory Johnston / Adobe Stock photo*

Right: A Canada goose and her goslings. *leekris / Adobe Stock photo*

MAMMALS

Small Mammals
- Badger
- Beaver
- House Mouse
- Masked Shrew
- Montane Vole
- Pocket Gopher
- Rabbit
- Bat
- Deer Mouse
- Least Chipmunk
- Meadow Vole
- Muskrat
- Porcupine
- Raccoon

Large Mammals
- Black Bear
- Cougar
- Lynx
- Mule Deer
- Wolf
- Grizzly Bear
- Coyote
- Moose
- Whitetail Deer

Herpetofauna
- Common Garter Snake
- Great Basin Spadefoot Toad
- Tree Frog
- Western Terrestrial Garter Snake
- Gopher Snake
- Long-toed Salamander
- Western Spadefoot Toad
- Western Toad

Above: A beaver and her kit.
Bailey Parsons / Adobe Stock photo

Right: A pocket gopher.
Kris / Adobe Stock photo

PRECIOUS GRASSLANDS   115

A coyote.
Harry Collins / Adobe Stock photo

MISCELLANEOUS HABITAT

Graminoids
Alfalfa
Alkali Saltgrass
Bluejoint Reedgrass
Common Timothy
Droopy Brome Grass
Giant Wildrye
Japanese Brome or Cheatgrass
Kentucky Bluegrass
Needle-and-thread
Pine Grass
Rough Fescue
Scratchgrass
Sough Grass
Spreading Needlegrass
Timber Oatgrass
Alkali Cordgrass
Bluebunch Wheatgrass
Chickweed
Crested Wheatgrass
Foxtail Barley
Green Foxtail
Junegrass
Little Meadow Foxtail
Nuttall's Alkaligrass
Porcupine Grass
Sandberg's Bluegrass
Slender Wheatgrass
Smooth or Hungarian Brome
Stiff Needlegrass

Shrubs
Bebb's Willow
Common Juniper
Creeping Juniper
Prickly Rose
Saskatoon
Smooth Sumac
Big Basin Sagebrush
Common Rabbitbush
Nootka Rose
Red Raspberry
Scouler's Willow
Squaw Current

Rushes
Bulrush

Seaside Arrowgrass
Wire Rush

Sedges
Beaked Sedge
Water Sedge

Trees
Cottonwood
Douglas Fir
Lodgepole Pine
Spruce
Trembling Aspen

Horsetails
Swamp Horsetail

Forbs
American Vetch
Arrow-leaved Balsamroot
Bearded Owl's Clover
Black Medic
Broad-leaf Plantain
Brown-eyed Susan
Canada Goldenrod
Canada Thistle
Celery-leaved Buttercup
Chickweed
Columbia Gromwell/
  Puccoon
Common Cattail
Common Chicory
Common Harebell
Common Peppergrass
Common Silverweed
Common Yarrow
Curlycup Gumweed
Cut-leaved Fleabane
Dandelion
Death Camas
Deervetch
Diffuse Knapweed
Douglas Catch Fly
Douglas Knotweed
Early Blue Violet

False Dandelion
Field Locoweed
Field Mint
Fine-leaved Fleabane
Fireweed
Flixweed
Graceful Cinquefoil
Great Mullein
Greater Plantain
Jerusalem-oak Goosefoot
Knotweed
Lamb's Quarter
Large-flowered Collomia
Long-leaved Fleabane
Low Pussytoes
Looseflower Lupine
Northern Bedstraw
Onion
Parsnip-flowered Umbrella
  Plant
Perennial Sow-thistle
Philadelphia Fleabane
Pineappleweed
Pink Microsteris
Prairie Sagebrush
Prickly Lettuce
Prickly Pear Cactus
Purple Avens
Red Clover
Red European Glasswort
Rosy Pussytoes
Russian Thistle
Sagebrush Mariposa Lily
Shaggy Fleabane
Shepherd's Purse
Shore Buttercup
Showy Milkweed
Silky Lupine
Silverleaf Phacelia
Silverweed
Small-leaved Forget-me-not

Spike-like Goldenrod
Spotted Knapweed
Star-flowered Solomon's Seal
Sticky Purple Crane's Bill
Strawberry-blite Goosefoot
Sulphur-flowered Umbrella
  Plant
Thompson's Paintbrush
Thread-leaved Fleabane
Timber Milk Vetch
Tufted White Prairie Aster
Water Smartweed
Wavy-leaved Thistle
Western Dock
Western Tansymustard
White Clover
White Sweet Clover
Wild Daisy
Wild Delphinium
Woolly Plantain
Yarrow
Yellow Rattle
Yellow Salsify (Oysterplant,
  Goat's Beard)
Yellow Sweet Clover

Following pages:
Grasslands
surround Nicola
Lake. *maxdigi /
Adobe Stock photo*

FOR YEARS, GARDNER was actively involved with Ducks Unlimited, an organization dedicated to conserving, restoring and managing wetlands and associated habitats for the benefit of North America's waterfowl.

Protecting the grasslands from weeds, fire and off-road vehicles is one of the important roles of Douglas Lake's range patrol. During the peak months for fly-fishing, when all the resorts and campsites are fully occupied, as many as 1,000 visitors can be using the recreational facilities at any given time. "To protect our grasslands, we keep visitors within the boundaries of our resorts, and they're generally good. Mostly, they are guys who want to go fly-fishing," says Wes Penny, who was hired in 2014 for security and range patrol.

A COURT BATTLE OVER LAND ACCESS

Trespassing on the ranch's private property has been an ongoing challenge that culminated in a multi-year court battle pitting Douglas Lake against the Nicola Valley Fish and Game Club and the Province of British Columbia.

"Nicola Valley Fish and Game Club claimed the public right to fish on Minnie and Stoney lakes," explains Gardner. "The case was complicated, involving old roads, new roads, old-survey private lots, fish farm licences, water licences, construction of dams, and private stocking of lakes since the late 1970s. One of the key issues was that a lake, as it was in the 1880s, was reserved to the Crown, but all the land around it, in every direction, is private. Also, a road did run across the lot, but a long way from the reserved portion of the lake.

"While investigating old files and documentation, it became clear that many different legal matters would be involved—old rights-of-way that are no longer used can still legally be public roads. The process to close them legally with the BC Ministry of Transportation and Infrastructure involves many steps."

To provide background, Gardner explains that in 1975, when Weyerhaeuser built a bypass road, at Douglas Lake's expense, the old section of the Pennask road, from inside the gate toward Pennask Lake, was no longer used by the travelling public and no longer maintained by the BC Ministry of Transportation. "In the 1980s, we locked the gate to protect our buildings and equipment, since we no longer had year-around staff there," says Gardner. "At the beginning, we received permission from the Ministry to lock the gate.

"Our success, along with that of Peter McVey [deceased former owner-operator of Corbett Lake Lodge] and others, since establishing a first-class fee fishery, only increased the public demand for access. Steps taken by Douglas Lake to close the old route drew the ire of the Nicola Valley Fish and Game Club. Attempts by the Ministry of Transportation to close the old road made things worse.

"The club filed a petition to have the old road declared public and have the lock removed. Both the provincial Ministry of Transportation and Douglas Lake felt that a petition was not the best way to deal with this very complicated and detail-specific matter, so we launched a counter claim in 2013, and the court case began.

"Our case was not heard until January 2017, in Kamloops, before Justice Joel Groves. After about a month, and then a site visit in May, we waited until December 2018 for the judgement. We were somewhat surprised that we effectively lost on every count. We could not understand how Justice Groves could ignore expert witness testimony, including survey details and hydrological information, and effectively give the public free access across our private lands.

"Due to the implications of what we thought was a flawed judgement, and the implications for many other ranchers and property owners, we decided to appeal, another expensive and slow process. The outcome of that appeal was that Douglas Lake Cattle Company was entitled to restrict access to Minnie Lake and Stoney Lake." The Nicola Valley Fish and Game Club appealed to the Supreme Court of Canada, which refused to hear that appeal.

Farming

As the early ranchers found out—the hard way—a particularly cold, snowy winter can result in cattle starving unless supplemental feed is provided. Fortunately, on the lower elevations of the Douglas Lake ranches, ample flat land exists for cultivation. The Thompson–Nicola Valley region of British Columbia is noted for having among the highest agricultural production in the province. Even further north, at the Alkali, Riske Creek and Circle S ranches, the fields are highly productive as long as water is available for irrigation. Located in semi-arid climate zones, all the ranches depend on irrigation.

Douglas Lake and Quilchena ranches have access to more water for irrigation than most agricultural operations in the Interior of BC. "We get water from our lakes and creeks," says farm boss Stewart Murray. "We have thirteen dams that create storage reservoirs, all for irrigation. Some have conservation licences for fish, but all are irrigation reservoirs first and foremost. We have twenty-four wells that we use every day for domestic water. Depending on the weather, we sometimes run out of water for irrigation by August, so in June we become stingy because we don't know what kind of a summer we're going to have."

Douglas Lake's northern operations, which are roughly 300 kilometres due north of the southern operations, are also in a semi-arid region, but the soil is drier and more porous, the weather is cooler, the winters are longer, and water for irrigation is a constant challenge.

In spring 2016, the water-pumping system that draws water from the Fraser River for irrigation at Deer Park Ranch was rebuilt. The original pump system was installed and improved by the two previous owners. The

Above: Chapperon after harvest season. *Courtesy Sherri Magee*

Left: The Divide Field at Quilchena being irrigated by pivot sprinklers. *Phil Braig*

FARMING   123

system features two pumps in the river, a concrete holding tank, and pipes and pumps to deliver the water to the pivot and wheel-line sprinkling systems 750 feet above the river. Without water from the Fraser River, the hay lands at Deer Park would not be as productive as they are now.

Water for Cotton Ranch comes from dams, Riske Creek and springs. At Alkali Lake and Circle S ranches, reservoir lakes feed into local creeks that provide sufficient water to grow crops. However, these water sources must be closely monitored, and the ranches are mindful of other users of the creeks. Range water is a concern because waterholes are dependent on spring runoff and precipitation.

Irrigation systems are expensive, but the more tons of feed that can be harvested per acre, the more cattle can be fed during the winter. Some fields are still flood-irrigated, but each year, the ranch adds more irrigation equipment. "In 1978, we got our first centre pivot," remembers Murray. "The year before, we had to cross swamps to cut the meadow, but when we put in the pivot, we levelled the land, and all of that increased our production."

Each year more pivot sprinklers are installed, some replacing wheel-line sprinklers, which are not as efficient. "On the pivots, the sprinkler rotators are closer to the ground, so there is less evaporation," explains Murray. "Less of the water turns to mist, and more water hits the ground. The newer pivots don't pressurize the water as much, so we spend less on pumping the water." Where the pivots are within the range of cell or Internet service, the speed and direction of the pivots can be controlled using a smartphone application. The only function that cannot be controlled by smartphone is the starting of the water pumps. Underground mainline pipes to the pivots are also continuously being added to improve efficiency of water usage.

Douglas Lake grows primarily alfalfa for hay and barley for silage, and in a typical year, harvesting starts in June and ends in September. "We have a lot of ground to cover in a hurry," says Murray. In total, including the northern operations, the ranch has more than 7,000 acres of irrigated land producing various crops for hay and silage.

"On Quilchena, we can get two to three cuts a year if the weather cooperates," says Murray. "At the Home Ranch and Chapperon, we can get two cuts, but at Norfolk we can only get one." The lowest elevation is at Quilchena, at about 2,200 feet above sea level, while the highest point of farmland is at Norfolk, at almost 3,500 feet. At the northern ranches, two cuts are typical, and three cuts are possible.

Seeding typically occurs between mid-April and mid-May, depending on the weather. That means harvesting can begin as early as June. "We're always fiddling with time," says Murray. "When we start off, the crops are at their peak nutritional value. By the time we get to some of the higher

Above: Harvesting at Chapperon. *Brent Gill*

Left: Grass silage being cut at Chapperon. *Brent Gill*

FARMING

elevations, the crops could be overripe. In a perfect world, we would have more machines and more people. We could start in June with both haying and silage crews going steady until October."

The difference in the elevations of the farms creates a natural schedule for seeding and harvesting at the southern operations. Delays can cause negative consequences that can ripple through the ranch's overall scheduling. Since 2017, one or more extreme weather events a year have been causing significant challenges.

In 2017, extreme weather started in January with heavier-than-usual snow and colder temperatures than had been experienced in decades. In the spring, heavy rain and snowmelt resulted in flooding that restricted access to some areas, prevented cattle from being moved, washed out roads and bridges, and delayed seeding of crops. Then, record-breaking temperatures, drought and wildfires hit with a vengeance. Fortunately, warm, dry weather continued into the fall, enabling a complete harvest.

"In 2017, we were late seeding at Quilchena, so we had to keep going, right up to Chapperon and Minnie Lake," says Murray. "We finished seeding about May 25 and on June 5 we took the first crop off at Quilchena using the same equipment. We finished haying September 10 and harvesting the rye

Corn growing at Quilchena.
Phil Braig

acres about the 15th. Then, we finished the corn by the end of September. Feeding started on October 5." Despite the late start, Murray and his crew successfully completed the harvest, thanks in part to the warm, dry fall.

"In 2017, we grew corn at Quilchena, and that's a crop we had never grown there," says Gardner. "We had installed some pivot sprinklers and ended up with a huge crop of about 20 to 23 tonnes of corn silage per acre. Our theory is that we might not need some of our higher elevation land that is least productive, where we get less tons of hay per acre. We're hoping to have higher production on fewer acres."

Having ample feed for winters was always a key strategy for Gardner. "Having feed in the pile is insurance," he says. "To get to the point that we were approaching—having some hay land that we didn't need to hay—was a good place to be."

The development of strains of corn that will grow successfully in cooler temperatures has made this crop possible at Quilchena. While corn is a finicky crop, the 2017 pilot project, planting about 100 acres of corn at Quilchena, was very successful. Timing was important. The corn had to be seeded, fertilized, soil-tested, sprayed for weeds, watered at the right time, and harvested at the right moisture content. The Quilchena area has the

The Grismer family, clockwise from bottom left: Emmett, Carson, Austin, Autumn, Emery, Heather, and dog Jupiter. *Courtesy Emery Grismer*

right growing conditions for the cooler-weather corn, but sufficient water availability is essential.

Winter 2017–18 was cold, snowy, long and especially challenging for the northern operations. In addition, smoke from the 2017 wildfires that had blanketed the area for eight weeks affected crop yields, resulting in less feed than usual for the winter. The snow, which typically melts by March 15, was still on the ground when wet, heavy snow fell at the beginning of April. Then, no precipitation at all fell until late June. The year ended well though, with ample hay and haylage.

"We have to be able to adapt," says Murray. "There's no point in fighting the weather. We have to adapt and balance what we can and can't do."

Weather is not the only change that has occurred in Murray's years at Douglas Lake. Times have changed, too. When Murray started working on the ranch in 1977, more people worked on the farm crew, equipment was smaller, and irrigation methods were less effective. "We worked six days a week, and sometimes seven days a week," he recalls. "Sometimes we got home at two in the morning. Now it's five days a week, and we have to make sure everybody has time off for a life besides working."

In August 2018, Emery Grismer joined Douglas Lake as assistant to farm boss Stewart Murray. Emery, his wife, Heather, and three of their four children (the oldest was attending university) moved to Douglas Lake from Saskatchewan.

"Emery is Stewart's right-hand man and has been rapidly proving to be a great addition to the farm operations," says Gardner. "He is an experienced truck driver, equipment operator, and farmer, and is able to supervise a large crew."

HEALTH AND SAFETY

Agriculture is considered a high-risk industry, and since Douglas Lake is the largest ranch in the province and at times has employed almost 200 people during the summer months, some in remote areas, the ranch attracts significant attention from safety officials.

"At first, agriculture was exempt from the Workers' Compensation Board's rules and regulations," Gardner remembers. "When WorkSafeBC (WCB at the time) started to study agriculture, the BC Cattlemen's Association asked me to sit on a committee, represented by all sectors of agriculture, to look at the issues. I went into this thinking we didn't need mandatory implementation, but after many meetings with each and every department of WCB, and learning first-hand about the carnage in agriculture, my opinion changed. We needed to embrace a better safety culture and move towards full implementation. It took a while before rules and regulations were implemented, but I believe our industry is in a much better place now, and I know our workers are.

"Douglas Lake ranches have regular committee meetings to deal with any safety issues and inspection reports. The culture is better, and most are buying in."

"The working guys are invested in safety," said Wes Penny when he was Douglas Lake's Occupational Health and Safety officer. "They want a safe place to work, even though it's more responsibility, more time, more paperwork, and this is a busy place. There are really good, hard-working people here who are committed to the ranch and understand that we can reduce risks with safe practices that might save their life."

FIRE SAFETY

Hot, dry weather can result in devastating wildfires. Two consecutive summers, 2017 and 2018, were the worst fire seasons on record in BC at the time.

Douglas Lake crews work to save grasslands during wildfires. *Gilitukha / Adobe Stock photo*

FARMING

129

While Douglas Lake's southern ranches survived the 2017 wildfires unscathed, its northern operations did not. At Riske Creek, four other ranches share Crown ranges in two different locations. Douglas Lake's deeded and leased lands were spared from the fire, but 70 per cent of the summer range and 90 per cent of the common fall pasture were burned. The ranch lost grass, a few historic buildings and some hay, but many neighbours' losses were much worse.

The wildfires moved quickly through the Riske Creek ranch grazing lands, burning everything in their path. Within weeks, though, after a heavy rain, the grass started to grow again.

While the southern ranches did not experience wildfires in 2017, prevention was high on everyone's mind. The ranches have some hydrants and some firefighting equipment, including a couple of horse trailers that have been converted to carry water tanks, pumps and hoses. With the nearest fire department located in Merritt, help could be an hour or more away. The neighbouring Upper Nicola Band also has firefighting equipment and training, so they and Douglas Lake back up each other. Ranch personnel have had to fight both building fires and wildfires in years past, and current employees train regularly to use the available firefighting equipment.

"In 1987, a disgruntled cook at the Home Ranch set the cookhouse on fire," Gardner recalls. "Because we had a fire truck and people trained to put out fires, we were able to save the cookhouse."

Some areas of the ranch, including some fishing camps, are remote and have no communications capabilities; it would require an hour's drive to deliver an emergency message in or out. That's one important reason why, during the dangerously dry 2017 summer season, some of those remote fishing camps were closed. The closures also minimized the risk of human-caused wildfires. At Stoney Lake, a powerful pump was installed to draw water from the lake to keep the vegetation around the lodge from burning during a fire. Although unnecessary during the 2017 fire season, the tops of buildings can now be sprinkled to protect against sparks from nearby fires.

Smoke from the 2017 fires drifted across most of BC, particularly its Interior regions, including Douglas Lake ranches. It affected not only outside workers but the crops, as well. Ash from the fires fell like snow, blanketing everything, and thick smoke blocked the sun, in some instances moderating the high temperatures.

The 2018 season was even more devastating, with more fires and more area burned, and Douglas Lake's southern operations saw significant damage in 2021. Due to the effects of global warming, larger and more frequent wildfires are anticipated in the future.

Policing

The Merritt RCMP provides policing services to Douglas Lake, but because assistance might be 45 minutes or more away, sometimes the ranch must deal with emergencies while waiting for help to arrive. "It was still dark at about four o'clock one morning when our phone rang," remembers Gardner. "It was Madeline, our cook, who lived by the store. She said someone was robbing the store. She could hear breaking glass. I told her to stay inside, and I quickly dressed, took my revolver and snuck up on the store. I was able to arrest the young, drunk individual, who had a drunk partner in our bunkhouse, where I placed the call to the RCMP on a pay phone. I placed the two individuals in the cookhouse washroom, under guard and in handcuffs (left over from my days as an auxiliary RCMP officer), while I had breakfast and waited for the RCMP to take the lads away."

MECHANICAL REPAIRS AND SOLUTIONS

With a diverse fleet of vehicles and equipment, ranging from highway trucks, pickups and all-terrain vehicles to lawnmowers, weed whackers and furnaces, Douglas Lake must be prepared to maintain everything in good working condition and repair breakdowns. "Although many employees have mechanical skills, we could not survive without our mechanics," says Gardner. "Jeff Simpson moved his family to Douglas Lake and became our mechanic. Jeff is very creative, can improvise and build just about anything. If anything breaks, Jeff can fix it. He built a screening plant for manure and gravel and a sawmill for cutting logs into timbers and decking for the ranch's many private bridges. The sawmill became known as Simpson Sawmills."

Timber Resources

Douglas Lake's main income stream is cattle. Second is timber. "With the amount of land we have, we have a lot of trees—probably $20 million of merchantable timber—on our property," Gardner said in 2019. "In our semi-arid climate, the trees don't grow as fast as in some other places, but they're still growing all the time.

"We want our forests to be sustainable. To me, that means we won't cut more than we have to. We never want the resource to be less than it was. We don't want to harvest more than the growth. If we harvest that way, we should be sustainable forever.

"Generally speaking, if we don't need to cut trees, we don't. This is about cash flow. If the cattle market is high, we let the trees grow. It's financial insurance, money in the bank. Some years we cut nothing. Any time we log, we make sure we have regrowth or plant trees.

"In most cases, we don't clear-cut. If we take out only about half the trees in an area, that gives the remaining trees more moisture, more sunlight, and the opportunity to grow faster. Cleaning up an area also improves grazing grasses and lowers fire hazards.

"This resource will be worth more in the future than it's worth now."

To achieve the goals for the ranch's timber resources, Douglas Lake has been working with Registered Professional Forester Marv Kempston of Westwood Fibre Resources for forty years.

After graduating with a degree in Forestry from UBC, Kempston worked with various forestry companies around the province before forming his own company in 1976. His core business is managing timber resources for farms and ranches. In 1980, Kempston started working with Douglas Lake to manage its timber resources, creating and ensuring a healthy, sustainable forest at each of the ranches.

"When I first started working with Douglas Lake, we inventoried the timber on the entire ranch," says Kempston. "We mapped out the possible blocks to log and [calculated] the degree of logging that would create a perpetual forest and avoid damage to bunchgrass areas.

"The logging operations have been conducted in forest areas where the only grass was pinegrass. Parking trucks or turning around logging trucks must be done away from bunchgrass areas. Also, we opened up the forest and spaced the trees far enough apart that sunlight could reach the ground. This allowed pinegrass—a valuable grazing forage—to grow again after having been shut out when the crowns of the trees grew together and blocked the sunlight.

"We selectively log by removing about half of the trees, so that the crowns of the trees have significant space between them. This allows sunlight to reach the ground again. We log across the diameter range, so that when we leave the forest there are still small, medium and large trees." The two main species of conifers on the Douglas Lake ranches are Douglas fir and lodgepole pine, but spruce can be found in moist areas.

An example of a selectively logged dry-belt Douglas fir stand in the Sabin area. *Jim Thrower, Westwood Fibre Resources*

"The higher quality logs are sold to sawmills and plywood plants, while the lower quality logs—pulp logs—are sold to whole-log chipping plants that produce woodchips for pulp mills," Kempston explains. "In the past, the slash piles developed from the logging operations have been burned in the fall of the year. Recently, the slash piles have been ground into a product called 'hog fuel' and sold to a power plant. This has eliminated the uncertainty of suitable conditions to conduct burning operations and eliminated the smoke from the fires."

Over the years, Kempston has dealt with a number of timber resource challenges at Douglas Lake. "At one point, a very significant infestation of dwarf mistletoe was causing a great deal of mortality in the lodgepole pine stands," he remembers. "We decided to clear-cut it, and that happened to be at the top of the market." Insects, such as tussock moths and bark beetles, and drought are ongoing challenges.

The acquisition of Quilchena and the northern ranches significantly increased timber resources, though on the Quilchena and Circle S properties most of the merchantable timber had already been harvested. "The trees on the northern ranches are subject to attack from Douglas-fir bark beetles and are in an area that has had significant fires in recent years," says Kempston. To manage these risks, the ranch and Kempston are planning extensive selection cutting on all those ranches.

According to Kempston, over time, the value of the trees and the value of grass per acre are roughly comparable, though the return on grass is annual, while the return on the timber might not be for eighty years. Most ranches do not have enough grass, he explained, and if the timber is not particularly valuable and treed areas are appropriate for grazing, the rancher might opt to cut the trees and grow grass. On the Douglas Lake ranches, the two are complementary, and logging is done in a way that protects and enhances the grasslands.

"It's important for me to know what the cowboys are doing," says Kempston. "We want to know that we made their job easier, that we repaired any fences we damaged, we piled and disposed of slash piles on a regular basis, we left new roads properly water-barred and suitable for access with horse trailers, and that we did not damage the bunchgrass areas."

Recreation Operations

Recreation started as a business for Douglas Lake during the Woodwards' ownership. Chunky Woodward was an avid hunter and fisher, and during his and his children's ownership years, they built fishing camps at Salmon and Stoney lakes. (Many of the lakes on the ranch have more than one name. For example, government maps refer to Salmon Lake, while some tourists and the Upper Nicola First Nation refer to the same lake as Fish Lake.) Many of Douglas Lake's natural lakes were dammed to create water reservoirs for irrigation and subsequently stocked with fish. Fishing is catch-and-release at Douglas Lake's stocked lakes. Provincially stocked lakes, such as Salmon Lake, have a limit of two fish per day. The ranch's lakes have attracted still-water fly-fishing enthusiasts for decades.

Visitors using Douglas Lake recreation facilities are restricted to roads and areas that will not damage the grasslands or interfere with the cattle operations. Agricultural bus tours had been allowed with the same restricted access. In 1991, Ingrid Boys, who had been a camp cook, approached Gardner with plans to expand the tour business. "In my year as a cow-camp cook, I'd learned about ranch history, ecosystems and operations, and was eager to share it," says Boys. "I think the idea appealed to Joe, in part, as it played into the ranch's 'offensive rather than defensive' strategy towards public interest in accessing the ranch. Offering access through tours would create a path to help reduce trespassing and damage to the fragile grasslands."

Right: A rainbow trout. *Rob / Adobe Stock photo*

Below: Fly-fishing season begins in mid-April and finishes in mid-October. *Brent Gill*

UPGRADING RECREATION facilities helped to retain the existing clientele and attract new customers. In 1991, Gardner built new facilities at Salmon Lake, including a new well, washhouse, store/residence, 11 cabins, and 24 RV sites; then, in the mid-1990s, built Stoney Lake Lodge. Twigg's Place, at Salmon Lake, was renovated in the early 2000s after Chunky Woodward's sister Mary "Twigg" White's family returned her lease to the ranch. Fishers who appreciated good fly-fishing and luxury full-service accommodations returned year after year.

In 1997, Gardner hired Douglas Lake's first recreation manager, Carlo Elstak. "We moved to the ranch in February 1997, during what was arguably one of the worst winters on record for BC," says Elstak, who is now general manager of Executive Hotels & Resorts at RCMP Pacific Region Training Centre in Chilliwack. "I learned very quickly that the tourist business is very seasonal, with a concentration in spring and fall for the excellent fly-fishing on the ranch's private lakes. We identified that during the hotter months, Stoney Lake Lodge, the jewel of the recreation accommodations, would make an excellent destination for European travellers, with a focus on an authentic Canadian ranch experience. We stepped up marketing efforts through new liaisons with international travel planners and tour operators and by attending travel trade shows. Before long, this started to pay off. Visitors from Britain, the Netherlands, and Germany were able to enjoy the lodge as part of their Western Canadian tours.

"An annual Fly-fishing Fiesta was established. Local guides, writers and TV sport-fishing personalities were invited to enjoy the ranch's extraordinary fly-fishing, which solidified Douglas Lake's position as the number one destination for private lake fishing in BC and the Pacific Northwest.

"The film industry also proved to be a lucrative opportunity, with several commercial and movie shoots providing yet another source of income, both for the ranch and for some of the cowboys who acted as extras. The feature film *The Snow Walker*, starring Barry Pepper, was partially shot at Douglas Lake and saw the introduction of reindeer onto the grassland for the first, and probably last, time ever.

"Through this diversification into new markets, a new era was launched for Douglas Lake as it positioned itself more prominently in the tourism business."

IN 2013, THE acquisition of Quilchena Ranch, with its historic hotel, golf course, and various RV and camping facilities, further expanded Douglas Lake's recreation business. Brent Gill, who joined Douglas Lake in 2011 as lodge manager at Stoney Lake, was promoted to recreation manager in 2013.

His wife, Dana, worked in central reservations and reception. At the time, thirty-five seasonal staff and fifteen off-season staff, mainly working at the hotel, were employed by the recreation division.

"We have recreation facilities from one end of the Douglas Lake property to the other," said Gill in 2017. "Between mid-April and mid-October each year, 15,000 to 17,000 people visit. Our regular seasonal guests come from as far away as New York and California, but generally most of our guests drive an average of four hours, with the biggest percentage from the Lower Mainland. Our facilities are booked months in advance and six of our properties have three-year waiting lists."

Douglas Lake's recreation business experienced significant challenges when damaging floods in the spring of 2017 resulted in the permanent closure of the Quilchena golf course; then, in the summer of that year and most following years, raging wildfires forced the closure of some or all of Douglas Lake's recreation facilities for varying lengths of time.

In 2020, the COVID-19 pandemic changed the business again. All rental properties were operated unstaffed, full-facility accommodation only, with the hotel and lodges rented to only one family or group at a time. "People were looking for this type of accommodation," says current recreation manager Yuki Sageishi. "Business started picking up again in 2021, and by the following year we were operating near full capacity."

In 2016, Yuki Sageishi and husband John Histed moved to Douglas Lake to manage the Salmon Lake Resort. Sageishi had worked in the hotel industry for more than a decade. Histed had been a hotel manager at Sun Peaks, BC, then general manager at a high-end hotel in Golden, BC. "Moving to Salmon Lake was a dream come true," says Histed. "Yuki and I love living close to nature and away from cities. Our dream was to run a remote fishing/hunting lodge, so Salmon Lake fits our lifestyles."

"This is what we both love doing, helping guests have a great stay, being outdoors, and talking fishing every chance we get," adds Sageishi. "We just love working for the ranch and living in this part of the country."

Left: Salmon Lake Resort offers a variety of accommodation types, including eleven cabins. Above: Stoney Lake Lodge is seasonally available as a full-facility rental. Below: Twigg's Place offers accommodation for eight. *Yuki Sageishi*

RECREATION OPERATIONS

Above: Yurt at Minnie Lake. Right: Quilchena Hotel. *Yuki Sageishi photos*

140 RECREATION OPERATIONS

Minnie Lake Ranch House. *Yuki Sageishi*

DOUGLAS LAKE RECREATION FACILITIES

- Salmon Lake Resort—Twigg's Place, rental house with accommodation for 8, 11 cabins, 32 RV sites, 30 campsites, 1 yurt, event tent
- Stoney Lake Lodge—8 guest bedrooms that can accommodate 15
- Minnie Lake—Minnie Lake Ranch House accommodates 8 people and has 2 yurts
- Quilchena Hotel—15-room historic hotel available for full-facility rental only (Quilchena general store is adjacent)
- Quilchena RV Park and Marina—40 RV sites on Nicola Lake
- Quilchena Creek long-stay RV facility—22 monthly rental sites
- Big Sabin Lake/ Little Sabin Lake—3 remote campsites
- Alleyne Lake—3 campsites
- Crater Lake—3 campsites
- Wasley Lake—2 yurts
- Mellin Lake—1 yurt
- Pikes Lake—day fishery
- Harry's Dam—day fishery
- Little Chapperon Lake—day fishery

"WITH ALL OUR facilities at full capacity, we can have upwards of a thousand people here," says Sageishi. "Most of our guests are repeat customers who have been coming here for thirty or forty years. We have a rebooking policy, so people can book their same spots for the next year. The majority of our guests are fishers, but in the summer months, families meet here, as well. At Salmon Lake, we have a kids' playground, horseshoe pits, a swimming pool, and kids can ride their bikes safely because the resort area is fenced. We have a fee-based boat launch and boat rentals."

The Douglas Lake Ranch Community

The distinctive white-and-red office at the Home Ranch.
Brent Gill

The Douglas Lake Home Ranch buildings, including homes, offices, general store, cookhouse, workshops, barns and church, are all painted white and have red roofs. White picket fences neatly tie the buildings together, giving it the cohesive appearance of a village, but it's the

people who live and work at Douglas Lake who truly create the ranch's sense of community. The owners and management of the day set the tone and provide the facilities, amenities, and the wherewithal to enable that sense of community to develop and strengthen. Despite the appeal of the ranching lifestyle and the vast beauty of the natural surroundings, without that strong sense of community the ranch's ability to attract and retain valuable employees and their families would be challenged.

THE COOKS

The cooks at Douglas Lake ranches play an important role, particularly in the lives of the single employees. The Home Ranch cookhouse is open 365 days of the year, at six in the morning, twelve noon, and six in the evening, to feed all the single employees three meals a day. It is also open for lunch for invited guests and others. In addition to the full-time cooks at the Home Ranch cookhouse, seasonal cooks work at the cow camps.

Ingrid Boys worked as a cow-camp cook in 1990. "Stan Jacobs hired me for spring turnout at Portland camp, working for Bill McDermid," says Boys. "Bill and his wife, Tereza, were good friends of mine, so I bunked with them for that twenty-three-day turnout. I was determined to be the best camp cook ever. I learned that beef and pie were expected staples, and at thirty-three years of age, I learned to make pie crust. I make a really good crust, a skill that still serves me today.

"I planned my first menu around beef. Chili with ground beef for the day we set up camp, followed by spaghetti and meat sauce the next day. I was surprised when I heard one of the guys say, 'Doesn't she ever cook beef?' Turns out that beef at a cow camp means great slabs of any cut, roasted, grilled or fried.

"I got up early, as I wanted to be sure that I had breakfast ready by 4:00, after the guys had jingled in and saddled their horses. By 4:30 or 4:45, they were off to gather the herd that Terry Milliken and his crew had turned out the day before. I often acted as traffic control when the crew drove the cow-calf pairs across Highway 5A. When I returned to the cookhouse, I prepared breakfast and then lunch for the fencing crew. The cowboys returned mid to late afternoon for a light meal (nobody told me I did not have to send them out with packed lunches). Dinner for both crews was at 6:00 p.m. That was a lot of cooking!

"Following Portland, I cooked for a couple of weeks at the Courtenay Lake camp for Terry Milliken's crew. My accommodation there was an old-fashioned prospector's tent, and the refrigeration was a bank of coolers with supplies from the Home Ranch general store.

"The crews disbanded after turnout, and I was offered the cook's job for Wendell Stoltzfus (aka Puck) at Hatheume camp. My wood cookstove was enormous. Our drinking water—which was delicious—was pulled from the creek a couple of hundred yards from my log cookhouse. Again, I had early mornings, but that was only to get the wood stove up to heat for a 5:00 a.m. breakfast.

"Gathering began in late autumn, along with the first snowfall of the year. From Hatheume, we moved down to Louis Corrals before Puck's crew returned to the Home Ranch. I finished my year as a cow-camp cook in mid-December, at Minnie Lake, working for Terry Milliken. I had a very small part in the story of Douglas Lake, but my time there was epic."

Mary Lavictoire has been a seasonal cook at the west end cow camp since May 2008 and handles the early and long hours cooking at three or four locations, with varying degrees of facilities. Kathie Lock and Sally Mergle have been cooking at the ranch alternating weeks since January 2015, living on the ranch while they're working.

Portland camp's Bill McDermid and Hatheume camp's Wendell Stoltzfus on a snowy Boxing Day at the Home Ranch.
Courtesy Ingrid Boys

Cookhouse Musical Chairs

"Mike Ferguson told me about his first visit to the Home Ranch cookhouse, in 1949," recalls Gardner. "He served up, went over to an empty seat, and asked, 'Can I sit here?' The response was 'Sit wherever you like.' So he did, and shortly thereafter someone else came along and said, 'You're in my seat.'

"People get used to sitting in certain seats. One of our cooks, Madeline, liked practical jokes, so I suggested for fun that she put out place markers, mixing up the seating. She and I loved it, but some did not see any humour in that. One employee actually called the owner, Chunky, in Vancouver to complain."

RANCH LIFESTYLE

The northern operations have fewer employees, scattered over a large area intersected by the Fraser River, so there is less interaction among the people living and working on the ranches there. Williams Lake is the nearest town, a half-hour to two-hour drive for employees at these ranches. A regular newsletter keeps everyone informed of what's happening throughout the company, and everyone is invited to the Christmas party, where service awards are presented.

The lifestyle holds an important appeal for people who choose to work on the ranch. Long-term employees, such as cowboy Jake Coutlee, former cow boss Stan Jacobs, farm boss Stewart Murray, and of course Gardner, stayed not just because of the work, but also because of the lifestyle.

"We can't speak about long-term employees without mentioning Jake Coutlee," Gardner said in 2019. "Jake has worked for Douglas Lake for more than sixty years and is a highly skilled cowboy. Others can learn from him just by paying attention and watching. He knows every fence, gate, horse, and where each horse came from and who first rode it.

"I once saw Jake watch a young, inexperienced cowboy mispair a calf and cow. Rather than correcting the cowboy, Jake waited until the cowboy left, then paired the correct calf and cow. Jake's a quiet, calm, deliberate person who doesn't want to lead but can certainly teach. He takes extreme pride in his work and is a valuable employee—that's an understatement!" Jake and his wife, Monica, reside on the Upper Nicola Reserve.

For Teresa Stewart-Brewer, who worked in the Douglas Lake office from 1994 until 2004, the last six years as Gardner's executive assistant, the ranch was more than a place to work. "Douglas Lake will always be home," she said. Teresa moved to the ranch with her parents, Don and Mary Stewart, in 1991.

Following pages:
Yuki Sageishi

Her father worked on the farm crew and her mother worked part-time as a cowboy. Teresa was in Grade 11 when her family moved to Douglas Lake, and since there wasn't a school at the ranch at the time, she, along with her sister and brother, rode the bus to Merritt schools. After graduation and a year at college, she returned to work at the Douglas Lake office. In 1999, she and ranch cowboy Bill Brewer were married at Salmon Lake Resort. "When Bill took a job in Saskatchewan, it was very hard for me to leave the ranch," Teresa said.

Children of families who work at Douglas Lake grow up and find jobs at the ranch, too. Cowboy Keith Smith, for example, lived at the ranch with his wife, Sandy, and when their children grew up, son Cody followed his father's footsteps and became a cowboy, while daughter Amanda worked in the ranch office.

Since Douglas Lake is more than 50 kilometres from Merritt and 90 kilometres from Kamloops, having accommodations available for employees is important. Most of the permanent full-time employees choose to live in the ranch housing provided at each of the ranches. In total, more than seventy houses, as well as bunkhouses, are available to employees.

DOUGLAS LAKE RANCH GENERAL STORE

Built in 1903, Douglas Lake's general store supplies ranch employees, neighbours and visitors with a wide selection of merchandise, including groceries, household supplies, stationery, hardware, clothing, cowboy gear, work accessories, videos, local arts and crafts, memorabilia, and, of course, Douglas Lake beef. Outside the store is a gas pump.

The distance to Merritt or Kamloops makes the ranch's general store not just a convenience but an essential service for employees. "Married people get married provisions," explains Neil Blackwell, who became the ranch's purchasing agent and store general manager in August 1994. "We used to have a dairy, vegetable gardens, and we gave employees a deal on beef. We had laying hens until the early 2000s, so eggs were free for staff until then too."

When Blackwell and his family moved onto the ranch, his wife, Gail, collected, hand-washed, candled, graded and packaged eggs, and if her husband couldn't deliver them, she did that too. "At one time, we had up to 4,000 birds," he remembers.

Today, the ranch has no dairy, vegetable gardens or hens, but families still enjoy discounts on those products and on cut-and-wrapped beef purchased through the general store.

Until 2019, a federal post office was located in the general store. "The general store is the community centre and was a post office for more than a

Douglas Lake general store. *Brent Gill*

hundred years," says Gardner, who was postmaster during the years he was general manager. "A company cannot be a postmaster, so successive ranch managers were postmasters. Everyone got their mail at the Douglas Lake post office." In 2019, the post office at the Douglas Lake general store was closed and moved to the Upper Nicola Reserve.

Leah Pockrant took over as storekeeper at the Home Ranch general store in 2015 and has since added new products and developed a giftware/artistry section of the store. "There are a lot of talented people in the area," she says. "I try to get the local people to showcase their work."

Pockrant started working at Douglas Lake in 2010 and went on to work at Salmon Lake Resort, Stoney Lake Lodge, Quilchena Hotel, and now the

Leah Pockrant, storekeeper at the Douglas Lake general store. *Courtesy Leah Pockrant*

general store. "Meeting new people is the best part of this job," says Pockrant, who got to know everyone who worked on the ranch when they came in for their mail. "I love this area. Once I'm off work, I ride, fish and hunt. I live in a small house with an amazing view. I like the outdoors, the peacefulness."

THE LEGENDARY FOLMER AND HIS MILK COWS

"I first met Folmer Nielsen when I was a summer student working at Norfolk," Gardner remembers. "At the time, Folmer was the milkman and also looked after the pigs and chickens. He lived in a small shack by himself, and the only time we saw him was in the cookhouse. He never talked to any of us.

"Fast forward to my coming back to the ranch and, lo and behold, Folmer was still there, living in the same shack at Norfolk.

"Neil, my predecessor, had decided that it made no sense for the cows to be at Norfolk when most of the milk was used at the Home Ranch. He also figured it was time to get some milking machines and modernize the dairy. Since Folmer had raised most of the cows and milked them seven days a week for so many years, Neil reasoned that he would welcome new facilities and better feed for the cows, and that he would enjoy being at the Home Ranch. Wrong. Folmer said no, he would not move.

"I agreed with Neil's thought that the dairy operation needed to be moved to the new facilities he had constructed at the Home Ranch. I reasoned that Folmer would come once we moved the cows. Wrong again. He would not move.

"We had created a monster, since we now needed one and a half milkmen and housing for both, as nobody would work seven days a week, twice a day. After a couple of years of various milkmen and various complaints, we found out we could buy the same amount of pasteurized milk and cream at half the cost of what we were producing. It was obvious that we needed to close the dairy.

"As for Folmer, he stayed at Norfolk, in his shack, and did flood irrigation and some haying work. However, he was a bit scary when it came to driving a tractor, because he had little or no experience and was stone deaf!

"Folmer had a pickup truck and would come down to the Home Ranch store weekly for groceries, which we provided. We installed a satellite TV in his shack, and he loved to watch the horse races. In the winter, we worried about him being alone up there, so I would go to Norfolk regularly to see how he was doing. One day, an RCMP officer friend came for a visit, and as we drove around the ranch we decided to check on Folmer. Folmer always had his door partially open, with a blanket hanging in the crack so that his cats, and later his chipmunks, could come and go without all the heat escaping. Banging on the front door, tapping on the windows, and turning on the truck siren did not raise him, so we went in. When we moved back the blanket separating Folmer's bedroom from the rest of the cabin, there he was, reading a book. When he saw us, he screamed. We retreated outside. Folmer finally came out and said, 'You thought I was dead.'

"Somewhere in the middle of all this, I asked Folmer if he had a will. He said no, he didn't need one. I convinced him he should have one, not necessarily for when he died, but in case he became incapacitated. I helped him find a lawyer, and he did make a will.

"In September 2014, at the age of ninety-two, Folmer died, in another shack we built for him when the first one burned down. He died of natural causes, and that's when I found out that he had named me executor and left everything to a cousin in Denmark whom he had met once on a trip there. Tracking her down and getting the funds to her was quite a project. How Folmer had accumulated $150,000 is a testament to how he saved, spending little."

GILBERT EYNON, CARETAKER OF THE PIGS

Gilbert Eynon arrived in Vancouver from England in the late 1950s, and through Chunky Woodward's brother-in-law, Robert White, ended up working on the ranch.

"Gilbert spoke with a distinguished English accent and, like Folmer, spent little of the money he earned from tending the garden at Chapperon, feeding the pigs, and doing various chores," remembers Gardner. "He couldn't drive and didn't know how to operate machinery. When the pig barn was closed, he continued to work in the large vegetable garden that fed the entire ranch. During the winter, he took the Greyhound bus from Merritt to San Miguel de Allende, Mexico, where he learned to speak Spanish.

"One year, Gilbert asked me for help regarding a letter he'd received. His spinster sister in England had died and left her estate to Gilbert, and he didn't know how to deal with it. I contacted Woodward's stores' buyer in England, who helped Gilbert sell his sister's home in Bexhill-on-Sea, in southeast England.

"Sometime later, Gilbert asked for more help when he received a cheque for $250,000 from the sale of the house. I suggested he start by depositing the money in his bank account. He gave me his bank book, which showed he'd saved some $150,000 while working for the ranch. His bank account was located in Vancouver, at the same branch where he had opened it upon his arrival in Canada, so I helped him deposit the cheque and also suggested he invest the funds rather than having that much money sitting in a savings account. I was able to help him with that, as well.

"A year or so later, Gilbert was diagnosed with a malignant brain tumor, so we moved him to the Home Ranch, where we could help take care of him. A young cook at the time, Lynne Whipple, would sit with Gilbert, make him tea, and drive him up to Chapperon, where he had worked most of his adult life. He enjoyed those drives. As his disease progressed, we moved him to palliative care in Vancouver.

"When Gilbert died, he left a third of his estate to a young Mexican boy who he had helped to learn English; a third to a cousin in England whom he had never met; and a third to Lynne, who by then had married and left the ranch."

Gardner and others looked after employees like Folmer and Gilbert, who worked for the ranch for decades and had no family to care for them in their final years. That's what people do for each other in a tight-knit community.

IN THE MOVIES

In 1982, Gardner was approached by a movie company wishing to film segments of *The Grey Fox* on the ranch. The movie was about notorious robber Billy Miner, also known as the Grey Fox, who was actually captured on the ranch in 1906. Miner had many aliases, including George Edwards, the name he used while working for Douglas Lake. He had been in prison in the United States for more than thirty years for robbing stagecoaches and trains but was perhaps best known for his escapes. After robbing a train in Oregon in 1903, Edwards turned up at the Douglas Lake Ranch, where he worked for Joseph Greaves briefly before robbing a CPR train with two accomplices. The three were eventually captured at Douglas Lake Home Ranch, but again Miner escaped. He was caught in Georgia after another train robbery and died in prison there in 1913.

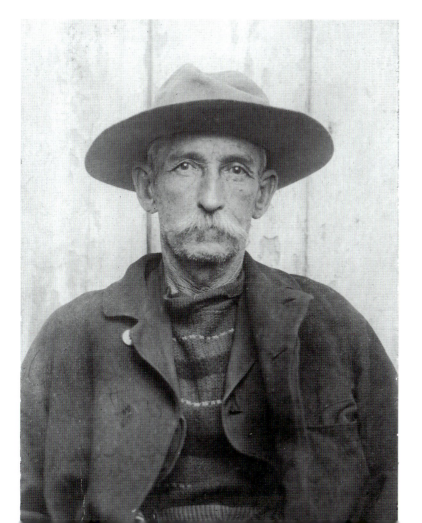

Below left: The notorious bandit Billy Miner briefly worked for Douglas Lake Ranch in 1906 before abandoning his duties to rob a CPR train with two accomplices. *Courtesy Kamloops Museum & Archives*

Below: Les Harris characterization by artist John Schnurrenberger. *Courtesy John Schnurrenberger*

"We agreed to the filming, and a large crew arrived and took over the Home Ranch," Gardner remembers. "They had a casting call for extras, and many of our staff were hired. One of our employees, our fencing foreman Les Harris, was particularly struck by this whole experience. He had a very distinctive look, which the movie people loved, and they managed to convince Les that he should go to Hollywood and have a portfolio done for future movies. Star-struck Les went off to Hollywood. While he was there, he phoned to advise me that he needed to stay longer. The following year, he returned to Hollywood. While I'm sure he had a good time, he never did get another acting role. In earlier years, Les had entered bull-riding events at local rodeos and had been severely banged up many times. On one occasion, he was almost killed."

Heeling Calves at the Big Meadow Branding, a painting by John Schnurrenberger.
Courtesy John Schnurrenberger

Swiss-born Canadian artist and cowboy enthusiast John Schnurrenberger painted numerous characterizations of Douglas Lake employees and scenes of life on the ranch. "The year 1976 was one that changed my life as an artist," Schnurrenberger says. "I had quit my job at the *Vancouver Sun* newspaper art department in the spring of 1974 and moved to Westwold to try to become a

At the Dry Farm cow camp, a painting by John Schnurrenberger. *Courtesy John Schnurrenberger*

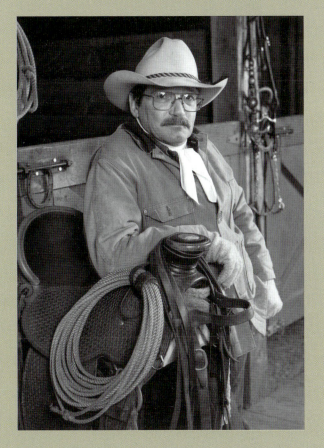

Canadian artist John Schnurrenberger cowboyed at Douglas Lake. *Courtesy John Schnurrenberger*

full-time self-employed Western artist. In the fall of 1976, I wrote a 'To Whom It May Concern' letter to Douglas Lake. Mike Ferguson answered and invited me to come up for one of the last fall drives. I was to bring my gear and meet him at the barn, where he would have a horse ready for me. I will forever be grateful to Mike, then others who took me under their wings and turned me into a half-assed hand." Over the years, Schnurrenberger attended brandings, turnouts, fall gatherings and drives, weanings, and other opportunities to learn about the cowboy life that he transferred to his paintings.

"Without the support of all these people, I could never have accomplished the kind of success I had in my forty-year career," says Schnurrenberger.

DOUGLAS LAKE SCHOOL

Historically, the Douglas Lake school was owned and operated by School District 58, based in Merritt. Douglas Lake donated a five-acre property, and the school board built a one-room school and a teacherage. When the population of children dropped below a certain number, the school was shut down. The ranch then rented or leased the facilities from the school board and operated its own school. The teacherage was set up as the community library, which previously had been in a room in the upstairs of the cookhouse.

Once the school district no longer operated the school, Gardner asked to have the property transferred back to Douglas Lake. "We struck a deal when we agreed to sponsor an annual scholarship of $1,500 to a graduating student for further education in agriculture," Gardner says. "The Douglas Lake Cattle Company scholarship continues to this day."

Chunky Woodward was the first ranch owner to recognize the need for a formal school for the children living there. "Chunky wanted families living and working on the ranch," remembers Carol Woodward. "Without a school, families would not live there, and without families, keeping good people would have been difficult. The working hours are long, and it's seven days a week."

"When I moved to the ranch, there was a one-room school with about twenty students, including a few from the reserve," remembers Sam Gardner, who moved onto the ranch in 1980 when she married Joe. "There was Kindergarten to Grade Six and one teacher, Griselda Evans, who was a friend from high school coincidentally. When families moved away and there weren't enough students, the school closed and the kids had to take a bus to Merritt. That was too hard on the little kids, so even more families moved."

In 1990, when Joe and Sam Gardner's daughter, Taylor, was ready to start school, there was no school at the ranch. Chunky Woodward was supportive of reopening the school, and when he died, his sons continued the support. "Sam supervised and did a lot of work upgrading the design and renovations of the school," says Gardner. "She was the first president of our Education Society, and Shirley Jacobs was the treasurer." The parents formed a school board and hired a teacher for Kindergarten and Grade One at the C.N. Woodward Elementary School.

The ranch paid the teacher's wages, and eventually the Merritt School District provided books and supplies. As the children progressed to the next grades, so too did the teacher, and when the students reached high school, another teacher was hired. "Megan Jacobs—whose father was the cow boss at the time—and I went all the way from Kindergarten to Grade 12 at the ranch school," says Taylor Gardner. "Our graduating class was two people, and that year there were ten students and two teachers in the school."

School being renovated, 1990.

Cameron Jacobs, Megan's brother, started Kindergarten two years after Taylor and Megan, and also completed Grade 12 at the school, graduating in 2005 with Neil Blackwell's daughter Amanda. "Especially in high school, I sometimes wished I were in a bigger school, but I got a good education because there was so much one-on-one time with the teacher," says Cameron. "Bruce Boulter was a particularly good teacher. He demanded a lot. He used to say, 'Not learning is not an option.' If you needed or wanted more time with him, he was always available. In the summer, he worked as part of the cowboy crew."

Sam Gardner made high school graduation a special occasion, not only for the Grade 12 graduates, but also for each and every student graduating to the next grade. Sam hosted a ceremony and celebration at the Big House, and Joe presented a silver belt buckle to the Grade 12 graduates. Hugh Magee, who always attended the graduation ceremonies, presented a pair of binoculars to each of the Grade 12 graduates.

Taylor Gardner

Taylor Gardner grew up on the Douglas Lake Ranch. "To have this whole place as a backyard to play in was the best," Taylor says. "I would go riding for hours, most of the time by myself. I'm not sure how I didn't give my parents grey hair, as it wasn't like today, when you can easily use a cell-phone to communicate. As far back as I can remember, I would jump on a horse as soon as school was done and be home in time for dinner. I grew up on the back of a horse. Sometimes I'd sneak out to ride. One time, when I was about five, I even fell asleep on my horse while he was eating, and I fell off. Horses and rodeo

Below: Taylor Gardner, Joe and Sam's daughter.

Above: Taylor riding Teal in her first Little Britches Rodeo, in Keremeos.

have always been my passions. I have been barrel-racing since I was six years old." Taylor is a fully carded Canadian Professional Rodeo Association and Women's Professional Rodeo Association competitor and continues to raise and train horses, and competes in Canada and the United States.

"One of the best things about growing up on the ranch was the valuable time I was able to spend with my parents, driving around the ranch together, riding, fishing and hunting, camping, Ski-Doo-ing, spending summers on the pontoon boat, and so much more."

Below: Taylor on Buster, Chunky Woodward's pack horse, 1988.

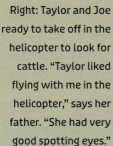

Right: Taylor and Joe ready to take off in the helicopter to look for cattle. "Taylor liked flying with me in the helicopter," says her father. "She had very good spotting eyes."

To raise funds for the operation of the school, Douglas Lake hosted an annual barn dance at the Quarter Horse barn. Tickets were sold in the surrounding communities for the fun evening of dinner and dancing, with entertainers such as Ian Tyson and Lisa Brokop. "These were sell-out events," says Gardner.

In 2011, when there were no longer enough students to keep the school open, Douglas Lake school was closed, but a school bus leaves from the ranch each morning and returns the students after school.

DOUGLAS LAKE COMMUNITY FUN

During the years when many families with young children lived at the ranch, parents organized sports and activities. "We played baseball, tennis, hockey; we had gymkhanas, swimming lessons and skeet-shooting," remembers Sam Gardner. "We were serious about our tennis tournaments. If the cowboys had to play, and they were out at camp, they'd come in jeans, cowboy hat and sometimes boots. But they made sure they got there to play."

Above: In 1993, Ian Tyson and Lisa Brokop entertained at the barn dance.

Right: Gymkhana, 1991.

THE DOUGLAS LAKE RANCH COMMUNITY

Left: The Tin Man, alias Joe Gardner, 1984.

Below: Hallowe'en party in the shop, 1989.

THE DOUGLAS LAKE RANCH COMMUNITY

In the spring of 1996, Gardner and Stan Jacobs purchased a 24-foot pontoon boat with a 40-horsepower motor for the ranch families and friends to use on Douglas Lake. "Many fun times were had pulling kids around the lake in tubes, fishing, and swimming behind the boat and off the beach," says Gardner. "The boat, which has a newer engine now, is still used by ranch families."

"In the winter, we made an ice rink on Sanctuary Lake," says Sam Gardner. "We got the tractor out there and got the snow off. We just had to make sure the ice was thick enough. We even made a kind of Zamboni using an oil barrel on skids. The floodlight at the end of the office lit up the ice. The kids and young cowboys formed a Douglas Lake hockey team, and when they were asked to play at an actual rink, it was the first time they saw boards. We just had snowbanks."

Special holidays were celebrated with parties for the kids and adults. There were Easter egg hunts and Hallowe'en parties, which included a hay-wagon ride, a bonfire and spectacular fireworks at the cookhouse, a haunted

Above: Santa arrives with presents for the children.

Right: Children sing in the church at Christmas, 2003.

THE DOUGLAS LAKE RANCH COMMUNITY

house that Sam Gardner created in the parts room of the shop, and a dance for children and adults, all in costume. Teresa Stewart-Brewer particularly remembered the hay-wagon rides at Hallowe'en, when the ranch children and some parents dressed up and went trick-or-treating from house to house. "Everyone knew everyone else, so the treats were more personal, many of them homemade," she said.

Christmas was especially fun for the kids because Santa would visit, and on his way he would talk with the excited children via two-way radio.

Each year, the Christmas party for adults is a dress event and awards are handed out for five, ten, twenty, twenty-five, thirty, thirty-five, and even forty years of service. During Gardner's tenure, the five-year award was an engraved knife in a scabbard; the ten-year award was an engraved leather belt; the fifteen-year award was a silver belt buckle; the twenty-year award was a trip to somewhere the individual wanted to go; and awards beyond that were customized to the employee.

Douglas Lake Neighbours

Douglas Lake's properties extend in various directions to meet neighbouring properties, which share common roads and face similar problems and challenges. Like all neighbours, they help each other and look out for one another.

The Douglas Lake Road, which runs south and west from Westwold for about 100 kilometres to Highway 5A, connects Douglas Lake with many neighbouring ranches. "In the mid-1980s, the BC Ministry of Transportation and Infrastructure began improvements to the Douglas Lake Road, from the Upper Nicola Reserve to the Spruce Field gate and, over the years, all the way to Big Meadow," says Gardner. "The windy dirt-and-gravel road was straightened and seal-coated. That was a huge improvement. The road through the reserve was still gravel, but some years later, that too was paved." The entire route, though, is still not paved.

LAUDER'S SPRINGBANK RANCH

Douglas Lake's closest neighbour is Lauder's Springbank Ranch, which started in 1876 when Joseph Dixon Lauder settled on the property west of the land John Douglas had preempted four years earlier.

Joseph's great-grandson John began managing the ranch in 1977. In 1980, John married Merritt veterinarian Dr. Jean Hansen. "Jean first came to Douglas Lake as a vet student," says Gardner. "Whenever we had pet issues, Jean was our go-to vet. We saw Jean and John socially and at events, including our Christmas parties."

After John and Jean retired, their son Ian took over as manager of Springbank Ranch.

GERARD GUICHON RANCH

The Gerard Guichon Ranch has always had a close relationship with Douglas Lake. Gerard and his wife, Ruth, were still the owners when Gardner was appointed general manager of Douglas Lake in 1979. "Gerard was famous for his work with the Canadian Cattlemen's Association," says Gardner. "He was credited with getting the family farm rollover provisions into the tax legislation, enabling tax-free rollovers of a family farm within the family."

In 1972, Gerard and Ruth's son Laurie returned to the family property after leaving his floatplane pilot job in Whitehorse, where he had met his Quebec-born wife, Judith. Seven years later, they purchased his parents' ranch.

"Laurie and I were both floatplane pilots, with each of us owning a Cessna 185, and we became good friends," says Gardner. "We flew all over the place for work and recreation, for example to Moses Lake, Washington State, for parts, and to Edmonton for a Cattlemen's meeting."

As friends with young families might do, Gardner and Guichon agreed to be each other's executors in the event that either of them died—of course, not thinking that would happen. "Unfortunately, one fine sunny day in 1999, while Laurie was on his way back to his ranch from a dentist appointment in Kamloops, he had a wreck on his Harley and died," recalls Gardner. "What a shock. We were unable to figure out what had happened because no other vehicle was involved, and he and his machine were both in good shape. What a tragedy.

"My executor role was simple, because Laurie had a will that was clear, and other than waiting for their four children to reach the age of maturity, there was little to do except sign some papers each year. Not too long after I signed off and was no longer an executor, Judy [Judith] called and asked to meet. Sam and I met Judy and eldest son, Mike. Judy asked if I would consent to being a blind trustee for everything she owned. We learned shortly thereafter that she had been named British Columbia's 29th lieutenant governor, a position she held for five and a half years. Being Judy's blind trustee also proved to be no real work, since her children Mike and Allison were looking after the ranch, and Tabitha and Darcy, also shareholders in the ranch, were not actively involved in the operations.

"While in Victoria for Judy's formal swearing-in reception at Government House, we met several members of the Government House Foundation, and I was subsequently asked to join the foundation as a trustee. I continue to enjoy that volunteer role, along with the new business and social connections. Since April 2018, when Judy finished her term as lieutenant governor, she has been back in the Nicola Valley, and we continue to see her, though in a much less formal way!"

CLEMITSON FAMILY RANCH

The Clemitson family, whose ranch is located in Westwold, has more than a century of history working for or with Douglas Lake. "Russell Clemitson was working on the Chapperon cowboy crew for Trevor [Thibeault] when he got wind that Meghan Bapty was working in the store at Salmon Lake Resort," Gardner remembers. "Russell knew Meghan's brother Nick. Meghan moved to the office at the Home Ranch after Salmon Lake closed for the season. Both Russell and Meghan left Douglas Lake in 2001 to go to the Clemitson ranch, where they were married in 2002. Their son Cache worked at Douglas Lake as a summertime cowboy for a few years, and then full-time for a year. Their other children, Cooper, Paisley and Griffin, haven't worked for Douglas Lake—yet! I remember Russell always being available to help us when we were short cowboys and driving truck for us when needed. The Clemitsons were always great neighbours."

"Trevor dropped me and my horse off on the road up to Gottfriedson Mountain and told me to ride up to the top—which seemed like a long way—to see if there were any heifers up there and then return to where he'd dropped me off," remembers Russell. "When I got to the top, I got off the horse to stretch my legs and ended up with my horse jumping out from under me and running back down the road. Whenever I got close, he ran further ahead. Five hours later, near the bottom, I got around him and was able to catch him. When I got on, he did the same thing. He blew up and dumped me, then ran off. Some hunters came along and, with a little help, I did get mounted and went and met Trevor, who gave me heck for being so late."

THE PENNASK LAKE FISHING AND GAME CLUB

The Pennask Lake Fishing and Game Club was started by Hawaiian pine-apple industry founder James Drummond Dole in the late 1920s. (Pennask Lake is south-southwest of the ranch.) Dole started filing land claims and buying land around Pennask Lake after he determined the lake was the perfect fishing spot for himself and his friends. He and fifty selected members created the exclusive club, which still exists as a private fly-fishing lodge at Pennask Lake.

"Pennask Lake boasts a large population of wild rainbow trout, and that's where the provincial government sources the genetics for their fish hatcheries," says Gardner.

"The deeded land around Pennask Lake has always been grazed by Douglas Lake cattle. The first formal lease on file required us to supply a fresh dairy cow each spring to provide milk and cream for the summer.

This deal morphed into us supplying weaner pigs each year to feed the table scraps to, with the resulting fat pig being butchered for staff each fall. The drive to Pennask is long and bumpy but was only six minutes in either my floatplane or helicopter. Sometimes I drove the piglets up to Pennask and sometimes I flew them up."

FIRST NATIONS NEIGHBOURS

"When I started as a summer student at Douglas Lake in 1963, I worked alongside many Upper Nicola First Nations people, like Ron Ned, whom I still know to this day," Gardner says. "Then, in 1973, I became the district agrologist for Indian and Northern Affairs in Kamloops. That led to extensive dealings and projects in the Nicola Valley and especially with the Upper Nicola Band.

"The Upper Nicola reserves total about 28,000 acres, and Douglas Lake employees pass through their reserves to go to town or anywhere else. With few exceptions, our dealings have been cordial, fair and respectful. Many of our employees over the years have been Band members, though much less now than in the past. Fewer Band members are interested in cattle or horses. The on-reserve membership of the Band has grown a lot. Housing and facilities have improved, with the big problem being lack of employment. While Douglas Lake used to have more than 200 employees, we have less than half that number now, resulting in fewer First Nations employees.

"In the early 1980s, our main swather guy was Joe Charters. He ended up knowing all our fields and did all the initial cutouts as he grew to know where all the ditches and rock piles were. Wayne McRae took over that task after Joe retired.

"Les Harris's fencing crew had Vern Charters, Archie Charters, Tracy Charters and Louis Holmes. Louie Stewart and his sons used to do contract haying for us and did fields like Howse Meadow and Stewart Field. His son Martin worked on the farm for Stewart Murray for many years as an equipment operator, until health issues forced him to retire. He and his wife, Nancy, live on the reserve. Former Chief Fred (Scotty) Holmes worked on the cowboy crew, as did Harold (Spook) McRae and Steve (Hyde) Archachan.

"The following Band members continue the good work for the ranch: Jake Coutlee, Brad Chillihitzia, Shane Charters, Alec (Bull) Chillihitzia, Wayne McRac, Roberta Dunn, Kyle Briggs, Joseph Munro, Ken Harry and Clifford McRae, to name a few.

"Obviously, Band members, both in the past and in the present, represent a significant portion of the total workforce of the Douglas Lake ranches and operating without them would be difficult."

The Evolution Continues

During the 135 years between 1884, when Douglas Lake Ranch started, and 2019, when Joe Gardner retired, the property and operations completely changed ownership and management relatively few times.

**Owners and Managers of Douglas Lake Ranch
1884–2019**

	Owner(s)	Manager
1884	Charles Beak Joseph Greaves Charles Thomson William Ward	Charles Beak (interim)
1885		Joseph Greaves
1892	Joseph Greaves Charles Thomson William Ward	Joseph Greaves
1910	William Ward	Frank Ward
1914	10 Ward siblings	Frank Ward
1940		Brian Chance
1950	Victor Spencer Bill Studdert	Brian Chance

1951	Victor Spencer Frank Ross	Brian Chance
1959	Charles Woodward John West	Brian Chance
1967		Neil Woolliams
1969	Charles Woodward	Neil Woolliams
1979		Joe Gardner
1990	Woodward siblings	Joe Gardner
1998	Bernie Ebbers	Joe Gardner
2004	Stan Kroenke	Joe Gardner
2019		Phil Braig

From the beginning, the numbers of acres and livestock were immense, and Douglas Lake was always foremost in its industry by just about any measure. Each owner and each manager added a distinct mark to the evolution of the ranch. For example, in the early years, Greaves assembled livestock and was intent on cornering the beef market in BC, while Beak assembled land and bred and sold Clydesdale draft horses. Woodward started a Quarter Horse breeding program and invested in irrigation systems, and during the forty years that Joe Gardner was Douglas Lake general manager, he implemented efficiencies and expanded the land and operations.

ON JULY 1, 2019, Phil Braig was promoted to general manager. "This has been the succession plan for many years, and it's now time for Phil to be the boss," said Gardner when the announcement was made. Gardner remained as vice-president of the company until his full retirement in 2021. He continues as a consultant to the owner.

Joe Gardner was general manager of the Douglas Lake Cattle Company for more than four decades, through four owners. He managed significant growth, improvements, and changes, and will forever be a legend at the ranch.

Quote from Stan Kroenke

"Douglas Lake is a crown jewel of ranching, due in no small part to Joe Gardner's dedication and leadership.

"For forty years, Joe gave his heart and soul to Douglas Lake Ranch. His passion for people, commitment to conservation and his reverence for ranches is inspiring.

"Joe's role in growing the business without compromising on the value and culture of Douglas Lake is remarkable. We'll always be grateful for Joe's tireless efforts to expand the scope and reach of Douglas Lake throughout Canada and North America.

"Joe set the gold standard for running and growing a major ranch operation efficiently and effectively.

"We will continue to do our part to ensure that this way of life that feeds and fuels so many communities will continue for generations.

"Thank you, Joe and Sam."

Stan Kroenke, owner
Douglas Lake Cattle Company

Opposite: Joe Gardner
and his wife, Sam.
Cathy Sloan

APPENDIX A
Alkali Lake Ranch

Alkali Lake Ranch dates back to 1861, when Herman Otto Bowe preempted 320 acres, where he farmed and offered food, drink and accommodation to those on their way to the Cariboo gold fields. He called his stopping house Paradise Valley, but those who stopped there often described the place as being "near the lake with the patch of alkali visible on the hillside," and eventually the location became known as Alkali Lake Ranch. Bowe acquired cattle to establish the first cattle ranch in the province, then partnered with John Koster in order to expand. Their sons Henry Koster and Johnny Bowe continued to operate the ranch until 1909, when they sold to Charles Namby Wynn Johnson, Chunky Woodward's grandfather. Johnson hired Jim Turner as cow boss. Turner had apprenticed under Douglas Lake's legendary cow boss Joe Coutlee.

Douglas Lake bought Alkali Lake Ranch in 2008.

In 1939, Johnson sold Alkali Lake Ranch to Mario von Riedemann, who had been a sugar beet producer and dairy owner in Austria. In the mid-1960s, Riedemann's youngest son, Martin, bought the ranch. He actively managed it until 1975, when he died in a boating accident on Alkali Lake. Two years later, Martin's widow sold the ranch to Doug and Marie Mervyn, from Kelowna. The Mervyns were pleased that Bill Twan, who had been managing the ranch since 1937, remained. Twan's son Bronc, who grew up on the Alkali Lake Ranch, remained as well, first as cow boss and then as manager after his father retired.

In 2008, Douglas Lake purchased Alkali Lake Ranch, the first of their acquisitions. The purchase not only significantly increased their operations but also started their northern operations.

Doug Mervyn was one of the two original owners and developers of Big White Ski Resort in Kelowna, BC. When they opened the ski hill in 1963, Big White boasted the longest T-bar in Canada.

In 1977, Mervyn was ready to move on from the ski business to become a rancher. His good friend Neil Woolliams, then general manager of Douglas Lake, told him that Alkali Lake Ranch, which was for sale at the time, was the best available ranch in the province. The Mervyns looked at other ranches but, after spending three days at Alkali, agreed with Woolliams and purchased the historic ranch. Alkali Lake Ranch was a larger venture than the Mervyns had initially set out to buy. Ranch manager Bill Twan and his son Bronc agreeing to stay on was an important factor in the Mervyns' decision to proceed with the purchase.

Mervyn had taken agriculture in high school, worked on three different ranches, and been a professional horse packer in Glacier National Park. To expand his knowledge, he read everything that he could find that was relevant to his new ranch, drew on the experience and successes of other ranchers, and joined ranching and cattle associations. As a result, in the thirty-one years that the Mervyns owned Alkali Lake Ranch, Doug and Bronc improved the grasslands, increased the capacity of the farmland, and built up the cattle herd. Besides helping her husband run the ranch, Marie became a founding member of the Fraser Basin Council.

APPENDIX B

Circle S Ranch

Circle S Ranch originated in the mid-1800s, when land was preempted by pioneers who were lured to the Cariboo by the gold rush but, for various reasons, chose instead to farm and ranch in the Dog Creek area. Among those whose preempted properties eventually became a part of Circle S Ranch were Le Comte de Isadore Versepuche, also known as Gaspard; Charles Brown; Moyse (Moses) Pigeon; Nels Gustafsen; Frederick Soues; William Liang Meason; William Holden; James Armes; Pierre Colin, later known as Peter Collins; John Gallagher; Thomas and William Patton; Raphael Valenzuela; and Bill Wright. These early ranchers sold and traded properties, but Joseph Smith Place was the first, in 1886, to start to consolidate these smaller land holdings into the larger operation that became known as Dog Creek Ranch.

The next major consolidation of ranches in the area began in 1931, when Colonel Victor Spencer (see page 26) bought several properties, including some of the Place family ranches, and created Diamond S Ranch, which owned ranches at Pavilion and Dog Creek Valley. Colonel Spencer continued to expand his holdings through the 1930s and 1940s, and in 1950 became an owner of Douglas Lake. Also in 1950, daughter Barbara Spencer became manager of her father's Dog Creek area ranches, renaming them Circle S. All of Spencer's ranches were sold after his death in 1960. Renee and Allerton Cushman, of Arizona, purchased Circle S Ranch, which then became a division of Sun Cattle Co. At the time, Circle S consisted of the Home Ranch at Dog Creek, Little Dog Ranch, Mountain Ranch, Pigeon Meadows Ranch, and Big Lake (Gustafsen) summer grazing range. Each of the individual ranches had living quarters and corrals.

Dog Creek Airport

Prior to World War II, commercial flights between Seattle and Alaska stopped at the Place family's Dog Creek property to refuel. During the War, the airfield was taken over by the Royal Canadian Air Force as a military base, with a radio transmitter station nearby. After the War, the Department of Transport took over operation of the runways. The airport was closed in 1960 when the Williams Lake Airport was opened, and in 1962, when the Cushmans purchased Circle S Ranch, the former airport land was a part of the property.

When US rancher Lyle James heard that Circle S Ranch was for sale in 1972, he decided to take the leap, despite the ranch being significantly larger than what he'd been looking to purchase.

Lyle and Mary James were both born in 1925 and raised on North Dakota ranches. After they married, they moved to Montana, then the following year bought a ranch in Bellevue, Idaho, which they sold two years later "at considerable profit." In 1948, they bought a ranch at Hot Springs, Montana, and for the next twenty-four years expanded their Garden Creek Angus Ranch and developed a small feedlot. Their first child, Linda, was born while they were in Idaho, and their subsequent seven children—Daryl, Janet, Don, Helen, Susan, Dale and Marilyn—were born while they were at Hot Springs.

"In 1961, Dad went on a trip to Calgary, landed at the auction yard, and that was when he fell in love with Canada," says daughter Helen. "Dad dreamed of owning a ranch in Canada." In 1972, Lyle and Mary purchased their dream ranch, Circle S.

"Buying Circle S was a huge step for Dad," says son Dale. "Compared to the 500 or so cows in Montana, Circle S was running about 600 head, but Dad saw the potential for that number to be much higher. Dad had purebred Angus in Montana, and he tried to maintain a purebred herd, but Circle S was so much bigger."

The move from Montana to Circle S was an ordeal because James chose to move all his cattle. At the Kingsgate border crossing, even though James had completed all the required tests on his livestock and had the required documentation and a customs broker, all the cattle and horses had to be unloaded and checked. Their one and only cattle liner was relatively small, and the multiple trips required to move all the livestock took five months, from September to January, driving more than 1,000 kilometres each way, through the Rogers Pass in the winter.

"On the last trip with the old Kenworth, the rear end of the truck gave out," says Dale. "The crown gear was cracked, it turned out, and had been the whole time he'd been driving back and forth."

From the outset, James faced challenges, but with his optimistic outlook, plenty of rangeland to feed the cattle, and neighbours who shared their valuable knowledge of the area, he and his family lived almost forty years on the Circle S Ranch.

"We had twenty foals a year, yearlings, and two-year-olds," Dale says about the large number of horses on their ranch. "We started breaking them at three years, and once they were broke we sold most of them. We were constantly training horses."

Dale remembers when 1,000 head of cattle were sold to the Gang Ranch in 1989: "Just counting them was a challenge. And then Dad was determined to build up the stock again. We had 600 first-calf heifers that next year, and we had to check them every two to three hours when they were calving." At its peak, Dale says Circle S had 3,000 head of cattle, including cows, bulls and yearlings, on 16,650 acres. "Dad loved stockyards, and he'd go and buy cattle and bring them back to the ranch to be branded before relocating them to a pasture or feedlot. We had cattle everywhere. We could hardly keep up. He increased range permits, too. There were ranges we didn't know. We would haul yearlings out to the range, which made gathering very difficult, because they didn't know where home was. It'd be February by the time we got them all in."

One year, when Marlboro cigarettes was shooting a television commercial at the Gang Ranch, Circle S was asked to provide a team of horses. Driving the horses to the Gang Ranch was worthy of a photo in *National Geographic*. Circle S also provided the setting for two films, *The 13th Warrior* (1997) and *The Thaw* (2008).

According to Dale James, his father's intent had always been to divide the Circle S assets among his children. In the 1980s, he formed a family trust that required the disposition of the assets of James Cattle Company by 2015. As the deadline approached, Lyle and Mary had to sell the ranch because it was unlikely that the beneficiary shareholders, their children, would be able to afford the taxes owing.

Circle S Ranch was sold to Douglas Lake Cattle Company in 2012.

APPENDIX C
Quilchena Ranch

Quilchena Ranch's roots originate in the mid-1850s, when the Guichon brothers—Charles, Laurent, Pierre and Joseph—left their home in Savoie, France, and travelled via California to the Cariboo gold fields. Recognizing that their future was not in gold itself, Charles, Laurent and Pierre established a supply and pack train service that catered to those who did believe that their fortune was to be made in gold. Rather than join his brothers, Joseph hired on with the largest pack train operator and the first beef supply company in BC.

Charles returned to France but continued to be a financial backer for his brothers. Pierre and Laurent bought cattle and started ranching at Mamit Lake, northwest of Merritt, and after Pierre died, Laurent and Joseph moved to property previously purchased at Chapperon Lake. When Charles Beak was on his quest to assemble land in 1882, the Guichon brothers' Chapperon Lake ranch was one of the acquisitions. With the proceeds from the sale, Laurent moved to the coast, and Joseph settled on his new Home Ranch at the mouth of the Nicola River, the start of what was to become Quilchena Ranch. Joseph Guichon's ranch became one of the largest stock operations in the region. In addition to cattle, he also raised Percherons and Thoroughbreds, which he sold to the North-West Mounted Police and Vancouver Mounted Police.

In 1904, Joseph Guichon purchased the land where the current Quilchena Hotel and general store are located. He built the hotel first, completing it in 1908, and the general store in 1912. With the railway rumoured to be coming through and Nicola Lake being a popular tourist destination, Guichon decided to build an elegant fifteen-room European-style hotel, which became a popular overnight stopover for stagecoaches travelling between the Coast and Kamloops. The railway never did pass through the area, and with reduced business resulting from World War I, the growing use of automobiles, and Prohibition, the hotel was forced to close in 1917. It wouldn't reopen until 1958.

In 1911, Guichon purchased Triangle Ranch—with a loan from Joseph Greaves—adding another 10,000 acres of land and 1,700 head of cattle. By 1918, when Guichon retired and his children, a son and two daughters, took over, the ranch included 40,000 acres of deeded land plus grazing leases on more than a half-million acres of Crown land. The eldest offspring, son Lawrence, became the manager of the ranch until 1947, when his son Gerard took over. Ten years later, the assets of the ranch were divided, and Guy Rose, another of Joseph's grandchildren, purchased the southern portion of the ranch, including the hotel and general store, and incorporated Quilchena Cattle Company. Gerard purchased the northern portion and incorporated Gerard Guichon Ranch Limited, which was subsequently taken over by Lawrence (Laurie) and his wife, Judith.

Guy and Hilde Rose, along with their children, Steve, Mike, Peter, Paul and Anne, expanded and improved their Quilchena Ranch. They reopened the hotel in 1958; started a three-hole golf course in 1963 and expanded it to nine holes in 1969; and built RV parks, campgrounds and fly-fishing facilities at the many lakes on the ranch. They also owned Quilchena Motors, a successful farm equipment dealership located behind the hotel. Steve Rose ran the dealership. For years, Douglas Lake bought International tractors and Heston haying equipment from Quilchena Motors.

In 2013, when Guy Rose's health declined and he was ready to sell Quilchena, Douglas Lake purchased the ranch.

APPENDIX D

Riske Creek Ranches

Both Cotton and Deer Park ranches were preempted by early settlers in the eastern Chilcotin. In 1868, L.W. Riske preempted the land that would become Cotton Ranch; and in 1873, Benjamin Franklin "Doc" English preempted land at the Deer Park Ranch location.

Cotton Ranch

In 1897, Riske sold his preempted land to Mortimer G. Drummond, who then started M.G. Drummond Ranch. About the same time, Robert Cecil Cotton embarked on a journey from his home in Hampton Court, Surrey, England, to the Drummond Ranch, where Cotton's father had arranged for his son to learn to be a rancher. "Mud pups" was the term used to describe these young men whose parents sent them to Western Canada and paid for them to learn to ranch.

When Drummond returned to England for a visit, he left his foreman, Charlie Moon, and Cotton to run the ranch. Moon had arrived from England in 1888, headed for the Cariboo gold rush, but along the way changed his mind and eventually ended up working at the M.G. Drummond Ranch.

Drummond returned from England and announced that he wanted to sell his ranch. Cotton bought it in 1907 and renamed it Cotton Ranch. He operated the ranch until 1945, when he sold it to John Wade, by then the owner of Chilco Ranch.

In 1961, Chilco Ranch was sold to John Minor, who died in a plane crash the following year. Shortly before his death, Minor had invited a number of partners and investors to join his ranching venture. "My father, Neil Harvie, Bryce Stringam and George McKimm, among others, invested," says Neil's son, Tim Harvie. "I'm not sure how Minor knew Dad, but probably through industry connections. My dad was always keen on adventures and challenges and loved ranching, so it would have been an easy sell to him."

Neil Harvie was the youngest son of oilman and philanthropist Eric Harvie, who held the mineral rights on properties where the first major Alberta oil discoveries were made. After graduating in Agriculture from the University of Alberta, Neil Harvie managed the family's Glenbow Ranch (now Glenbow Ranch Provincial Park), which he inherited from his father, then expanded his holdings in Alberta and British Columbia.

"When Minor died, my dad effectively became the managing partner, partly to protect his investment, but also to see Minor's dream fulfilled," says

Neil Harvie at a Riske Creek ranch in the 1960s. *Courtesy Harvie family*

Tim. "He hired Tom Livingston as ranch manager to keep the operations going and Lynn Bonner to manage the farming operations."

"Dad spent about a week a month at Chilco managing the business," says Tim Harvie. "He became familiar with the area and the vast properties involved. Dad realized Cotton Ranch was the jewel of Chilco because of its location, water supply, grazing leases and permits, and decided to buy it." That was the start of Riske Creek Ranching. To secure a feed supply, Harvie also bought Deer Creek Ranch, most of which was in irrigated hay production. Harvie asked Bonner, whose family were long-time residents of the Chilcotin area, to develop Deer Creek Ranch. The cows were herded along Highway 20, which wasn't paved at the time, to Deer Creek each fall to winter at the feed supply, then moved home in the spring to calve.

Lynn Bonner married Myrna Huffman, making Grant Huffman his brother-in-law. Huffman had worked as a ranch hand at Alkali Lake Ranch before starting in Agriculture at the University of British Columbia, where he and Joe Gardner became friends. During summer breaks, he worked at Chilco Ranch, where Lynn Bonner was manager. Huffman finished his master's degree and started his doctorate, but when Harvie proposed that he and Bonner manage the Riske Creek ranches, Huffman left university, and in 1972 he and Bonner formed a ranch management company, Two Rivers Ranching Ltd., and he moved to manage Cotton Ranch.

Junction Sheep Range Provincial Park

The junction of the Chilcotin and Fraser rivers has spectacular cliffs, hoodoos, river valleys and grassland benches. For generations, this has been the home of the world's largest population of non-migratory California bighorn sheep. In 1973, when Riske Creek Ranching purchased grazing ranges in the area from the Gang Ranch, the provincial government proposed swapping the lots they felt were important as sheep habitat for lots at a higher elevation, considered of less value to the sheep. At that time, the provincial government established an 11,300-acre wildlife reserve to protect the sheep and their natural grassland habitat. In 1987, the reserve was designated as the Junction Wildlife Management Area. In 1995, the Province of BC re-designated the area Junction Sheep Range Provincial Park.

Deer Park Ranch

In 1886, W.J. Drummond purchased the 320 acres that "Doc" English had preempted in 1873. There is no conclusive evidence that W.J. Drummond and M.G. Drummond were related, but the facts that both were from England and both bought ranches so close together in the Chilcotin suggest that they might have been related.

W.J. Drummond's ranch became known as Deer Park Ranch because of the large number of deer that roamed the ranges. In 1890, Drummond sold his ranch to Herbert Davis, who subsequently sold to Charlie Moon when Moon and his family returned from England in 1907. Moon bought neighbouring ranches and homestead properties and built a cattle empire. Before his death, Moon divided the ranches among his three sons, with Jack Moon acquiring Deer Park Ranch. When Jack died, his son Rob continued there until 1981.

"When Deer Park Ranch became available in 1981, Dad bought it to replace Deer Creek Ranch, because its proximity to hay production blended perfectly with Cotton Ranch," says Tim Harvie. Once Highway 20 was paved and traffic increased, the spring and fall cattle drives between Cotton and Deer Creek ranches became problematic.

Bonner relocated to Deer Park Ranch, while his sister Trena and her husband, Wayne Plummer, took over at Junction. Two Rivers Ranching became partners with Harvie in Riske Creek Ranching.

Neil Harvie, Lynn Bonner and Grant Huffman were owners of Riske Creek Ranching. *Courtesy Harvie family*

Riske Creek Ranching

In 1999, when Neil Harvie died, his four children inherited their father's ownership in Riske Creek Ranching. "I was already a director of the company and had been joining Dad on his trips to the ranch, so was very familiar with the ranch and the people [Grant Huffman and Lynn Bonner] involved," says son Tim.

In 2003, when Bonner's wife became ill, he sold his shares in Riske Creek to Huffman. He and his wife retired on a quarter section they purchased at Riske Creek. Myrna Bonner died in 2007, and two years later Bonner married Kathy Lauriente. They still live on their Riske Creek property, where Lynn remains active with welding, shop work, blacksmithing, riding his Harley trike, and flying his Supercub.

Huffman initially managed both Cotton and Deer Park ranches. When his son Cuyler, who had been working at the ranches during summers, graduated from Olds College, he took over as manager of Deer Park Ranch.

During his years of ranching, Huffman was actively involved in cattle associations, including being president of BC Cattlemen's Association from 1984 to 1986.

When the Harvie family decided to sell Riske Creek Ranching, Huffman was interested in buying the remaining shares but was beyond the age of wanting to take on such a large debt. With the sale of Riske Creek Ranching to Douglas Lake Cattle Company, Huffman retired.

APPENDIX E

Gang Ranch

In 2022, Douglas Lake Cattle Company acquired the historic Gang Ranch, which added another 28,000 acres of deeded land and 700,000 acres of Crown grazing land.

The Gang Ranch, originally known as the Canadian Ranching Company, began in 1863 when Jerome and Thaddeus Harper from West Virginia obtained a Crown grant on 160 acres along Gaspard Creek.

The Harpers, who aspired to build the largest cattle ranching operation in the world, leased and acquired increasingly more prime grazing land west of the Fraser River and through the Interior of British Columbia. At one time, the Harper brothers were the largest landowners in BC, and the Gang Ranch included land at Canoe Creek, Kelly Ranch, Harper Ranch, Perry Ranch, 57 Mile, Sullivan Pastures, Meadow Lake, and Crows Bar.

The origin of the name "Gang Ranch" is uncertain but refers to either the large "gangs of workers" required to establish the irrigation ditches or the "gang plow," which the ranch used before anyone else in the area.

Thaddeus continued to operate the Gang Ranch after Jerome died in 1874. In 1876, when the price of cattle dropped in BC, Thaddeus and a crew of ten cowboys began the legendary "longest cattle drive in BC history." He started with 800 head in BC and bought more along the way to his intended destination of Chicago, where he believed he could sell the cattle at significantly higher prices than he could in BC, even after factoring in all his costs. They wintered the cattle near Walla Walla, Washington. By spring, prices in the Chicago market had dropped drastically, so Harper grazed the cattle in Idaho, where feed and water were plentiful. Meanwhile, drought in California caused the death of thousands of cattle in that state. Eighteen months after leaving Gang Ranch, Harper sold 1,200 head of fattened cattle in the booming San Francisco market, at even higher prices than he had originally hoped to sell for in Chicago. He later shipped additional cattle to San Francisco, but that wouldn't be enough to prevent financial hardship in the decade that followed.

In 1888, already having sold some of his properties, Thaddeus sold the Gang Ranch's remaining holdings at Canoe Creek, Clinton, Kamloops and Cache Creek to his partners in Western Canadian Ranching Company (WCRC), headed by London publisher Thomas Dixon Galpin. Galpin sent sons-in-law Cuyler Holland and James Douglas Prentice to WCRC's

Gang Ranch general store and post office.

Canadian headquarters in Victoria, BC, to manage the Gang Ranch and the company's other ranches in Canada and the US.

During the years that the Gang Ranch was owned by WCRC and managed by a succession of managers, improvements were made to the facilities, including the building of a large "rest house," which became known as "The Big House" and was the main residence of the English managers and overnight guests. A barn, store, cookhouse, bunkhouse and infrastructure were built, and crops and cattle improved.

In the late 1920s, as debt increased and cattle prices and land values plummeted, the Gang Ranch once again found itself in financial difficulty. The situation worsened during the Depression of the 1930s, and WCRC struggled for survival. The late 1930s brought even more challenges, and by the end of the decade, WCRC could no longer survive its heavy debt and high operating costs.

In 1948, WCRC sold the Gang Ranch to sea captain Bill Studdert and his business partner, livestock trader Floyd Skelton. Studdert and Skelton sold the majority of the cattle to pay for the ranch. Studdert managed the ranch, spending as little as possible and allowing the ranch to deteriorate.

In August 1950, two years after buying the Gang Ranch, Studdert joined Victor Spencer in purchasing Douglas Lake Cattle Company. Months later, they invited Frank Mackenzie Ross to join their Douglas Lake venture, and when Ross refused to be in partnership with Studdert, Spencer chose Ross over Studdert. By April 1951, Studdert was no longer an owner of Douglas Lake.

In 1958, Skelton hired a new manager, Melvin Sidwell, to replace Studdert at the Gang Ranch. Sidwell improved the cattle herd, irrigation and hay production; updated the machinery; added new infrastructure; and brought electricity by generator to the ranch. Sidwell left in 1963 and was replaced by Wayne Robinson, who managed for three years and during that time achieved the highest levels of production the ranch had yet seen. In 1966, Sidwell returned and continued to improve machinery, land, livestock and buildings.

In 1971, Studdert died, which resulted in the sale of his interest in the Gang Ranch. Seven years later, Skelton sold Gang Ranch to the Alsagar family, from Alberta and Saskatchewan, for $4.2 million. At the time, Gang Ranch included 38,000 deeded acres and 700,000 acres under lease. For the

Gang Ranch. *Courtesy Douglas Lake Ranch*

first time, the Gang Ranch was owned by Canadians, but after five years of family and financial challenges, the ranch went into receivership.

In June 1984, the Gang Ranch was sold to Melvin Nelson, from High River, Alberta, but within months he sold to an investor group headed by John Rudiger. Rudiger's group included a Pennsylvania-based corporate investor, B.S.A. Investors Ltd., backed by sheik Ibrahim Muhammad Afandi, from Saudi Arabia. B.S.A. bought out its partners and became the sole owner of Gang Ranch in 1987.

In April 1990, when Larry and Bev Ramstad began managing Gang Ranch, buildings were in disrepair, fences were nonexistent, the irrigation system and hayfields needed attention, and the cow herd was in poor shape, as was the horse herd. The Ramstads built fences, repaired and upgraded buildings and irrigation systems, improved hay production, and established a breeding program for a top-quality herd of black Canadian Angus cattle and another to produce superior Quarter Horses.

Douglas Lake's acquisition of the Gang Ranch in 2022 brought the two most widely recognized, historic cattle ranches in British Columbia under the same ownership.

Bibliography

Bonner, Veera; Bliss, Irene E. Litterick, Hazel H. *Chilcotin: Preserving Pioneer Memories*. Surrey: Heritage House Publishing Company Ltd., 1995, reprint 2005.

Carroll, Campbell. *Three Bar: The Story of Douglas Lake*. Vancouver: Mitchell Press Limited, 1958.

Logan, Don. *Dog Creek: 100 Years*. Victoria: Trafford Publishing, 2007.

Mather, Ken. *A Short History of Cattle Ranching in the Cariboo*. Victoria: Royal BC Museum, 2002.

Roger, Gertrude Minor. *Lady Rancher*. Surrey: Hancock House Publishers Ltd., 1981, reprint 1988, 1991.

Stangoe, Irene. *Cariboo–Chilcotin: Pioneer People and Places*. Surrey: Heritage House Publishing Company Ltd., 1994.

Woolliams, Nina G. *Cattle Ranch: The Story of the Douglas Lake Cattle Company*. Vancouver: Douglas & McIntyre, 1979.

Periodicals
Beef in BC
Bridge River Lillooet News
Williams Lake Tribune

Archives and Online Reference Sources
BC Archives — Royal BC Museum
British Columbia Cattlemen's Association
Canada: A Country by Consent
City of Vancouver Archives

Index

13th Warrior, The, 176

A

Afandi, Ibrahim Muhammad, 187
Alkali Lake Ranch, 29, 90–91, 124, 172–73
Alleyne Lake campsites, 141
Alsagar family, 186
American Quarter Horse Association, 53
American white pelican, 111–12
Anthony, Fin, 43
Archachan, Steve (Hyde), 167
Armes, James, 174
Arnell, Dale, 61, 78–79
Arnell, Magy, 79

B

Bapty, Nick, 166
barcode ear tags, 104
barn dance, 160
Barnett, Duncan, 64
Barnett, Jane, 64
BC Cattlemen's Association, 128, 183
BC Livestock auction, 106
BC Livestock Producers Co-operative Association, 70
BC Ministry of Transportation, 120, 164
BC Verified beef, 106
Beak, Charles Miles, 19, 168–69, 177
Bell-Irving, Henry "Budge", 71–72
Bell-Irving, Nancy, 71
Big Lake (Gustafsen) grazing range, 174
Big Meadow, 67, 107, 154, 164
Big Sabin Lake, 141
Blackwell, Amanda, 148, 157
Blackwell, Gail, 81, 148
Blackwell, Neil, 81, 148, 151
bluebunch wheatgrass, 89, 111, 116
Bonner, Kathy (Lauriente), 182
Bonner, Lynn, 180–82
Bonner, Myrna (Huffman), 180, 182
Boss (The) - Clydesdale stallion, 21
Boulter, Bruce, 157
bovine anaplasmosis, 104
bovine tuberculosis, 105
Bowe, Herman Otto, 172
Bowe, Johnny, 172
Boys, Ingrid, 135, 143
Braig, Phil, 7, 97–99, 169
Brewer, Bill, 148
Briggs, Kyle, 167
British Cattle Breeders, 70
British Columbia Land Ordinance Act, 15
Brokop, Lisa, 160

Brown, Charles, 174
Bryson's Ranch, 26
B.S.A. Investors Ltd., 187
Burke, Tony, 55

C

California Big Horn sheep, 181
Canadian Cattlemen's Association (Canadian Cattle Association), 56, 105, 165
Canadian Cattlemen's Foundation (Canadian Cattle Foundation), 70
Canadian Cutting Horse Association, 38
Canadian Food Inspection Agency (CFIA), 104
Canadian Pacific Railway (CPR), 16
Canadian Satellite Livestock Auction, 106
Cariboo Gold Rush, 15–16, 19, 174, 179
Carnegie Hero Fund Commission, 58
Castonguay, Peter and Marina, 63
cattle
    breeds, 101
    calving, 102, 107, 176
    marketing, 105–6
Chance, Brian Kestevan de Peyster, 25
Chapman's Cabin, 57
Chapperon Lake, 19, 21, 61, 63, 177
Chapperon Lake Ranch, 177
Charters, Archie, 167
Charters, Joe, 167
Charters, Shane, 167
Charters, Tracy, 167
Charters, Vern, 167
Chilco Ranch, 179–80
Chillihitzia, Alec (Bull), 167
Chillihitzia, Brad, 167
Circle S Ranch, 26, 91–93, 124, 174–76
Clemitson, Cache, 166
Clemitson, Cooper, 166
Clemitson, Griffin, 166
Clemitson, Meghan (Bapty), 166
Clemitson, Paisley, 166
Clemitson Ranch, 166
Clemitson, Russell, 166
C.N. Woodward Elementary School, 156
Colin, Pierre (aka Peter Collins), 174
cooks, 143–45
Coquihalla Connector (Okanagan Connector, Highway 97C), 76–77
Corbett Lake Lodge, 72, 120
corn, 126–28
Cotton, Robert Cecil, 179
Cotton Ranch, 124, 179–81
Courtenay Lake camp, 144
Coutlee, Jake, 63, 145
Coutlee, Joe, 63, 172
Coutlee, Monica, 145
COVID-19, 138

Crater Lake campsites, 141
Crown grazing ranges, 89–90
Cunliffe, Dave, 81
Cushman, Allerton, 174
Cushman, Renee, 174

D

dairy, 151
Davis, Herbert, 181
Deer Creek Ranch, 180–81
deer farm, 77–78
Deer Park Ranch, 93, 122, 179, 181, 183
Diamond S Ranch, 26, 174
Dinsdale, Harold, 55
Dinsdale, Leah, 55
Dog Creek, 26, 91, 174
Dog Creek Airport, 174
Dog Creek Ranch, 174
Dole, James Drummond, 166
Douglas, John, 14–15, 19, 164
Douglas Lake Amendment, 76
Douglas Lake beef 105
Douglas Lake Cattle Company scholarship, 156
Douglas Lake church, 84, 86
Douglas Lake Equipment, 88–89
Douglas Lake general store, 142, 148–150
Drummond, Mortimer G., 179
Drummond Ranch, 179
Drummond, W.J., 181
Dry Farm, 33, 55, 63, 104, 155
Ducks Unlimited, 120
Dunn, Roberta, 167
Dynneson, Leann, 55
Dynneson, Tom "Rock Creek", 54–55

E

Earlscourt Ranch, 26
ear-tag program, 79, 104
Ebbers, Bernard, 83–87
Ebbers, Kristie (Webb), 84–86
Ednoste, Tom "Forty", 55
Elstak, Carlo , 137
Empire of Grass, 12
English, Benjamin Franklin "Doc", 179, 181
English Bridge, 54–56, 61–63, 106
Evans, Griselda, 156
extreme weather, 23, 126
Eynon, Gilbert, 152

F

Ferguson, Debbie, 35
Ferguson, Mike, 35, 46, 63–65, 70–71, 105, 107, 145, 155
film industry, 137
fire safety, 120–30
Fish Lake, 135

189

fishing, 135–137
floods, 138
Fly-fishing Fiesta, 137
Forbes, Malcolm, 73–74
Fort Macleod Auctions, 106
Fraser River gold rush, 15
Frelick, Gary, 89

G
Gabara, Kathy (Kate Hendrickson), 55
Gallagher, John, 174
Galpin, Thomas Dixon, 184
Gang Ranch, 63, 175–76, 181, 184–87
Gardner, Joe, 43–47, 55–56, 60, 70, 80–82, 91, 159, 161, 169–71
Gardner, Joseph, 43
Gardner, Sam (Sandra Jean Stein), 46, 56, 73, 156–57, 170–71
Gardner, Taylor, 87, 156–59
Gerard Guichon Ranch, 165, 178
Gill, Brent, 137–38
Gill, Dana, 96, 138
Glenbow Ranch, 179
Glenbow Ranch Provincial Park, 179
Government House Foundation, 165
Grabowsky, Kelvin, 63
grasshoppers, 21, 24, 26
grasslands, 89–91, 101–2, 104, 111–12, 120
Great Depression, 25
Greaves, Joseph Blackbourne, 16–23, 153, 168
Grey Fox, The, 52, 153
Grismer, Austin, 127
Grismer, Autumn, 127
Grismer, Carson, 127
Grismer, Emery, 127–28
Grismer, Emmett, 127
Grismer, Heather, 127–28
Grosvenor, Major General Gerald Cavendish, 71
Grosvenor, Natalia, 71
Groves, Justice Joel, 121
Guichon, Allison, 165
Guichon, Charles, 177
Guichon, Darcy
Guichon, Gerard, 165, 178
Guichon, Joseph, 177
Guichon, Judith, 165, 178
Guichon, Laurent, 177
Guichon, Lawrence (Laurie), 177
Guichon, Pierre, 177
Guichon, Ruth, 165
Guichon, Tabitha, 165
Gustafsen Lake (Big Lake), 174
Gustafsen, Nels, 174

H
Hansen, Amanda, 72–73

Hansen, Rick, 72–73
Harper, Jerome, 184
Harper, Thaddeus, 18, 184
Harris, Les, 153–54, 167
Harris, Madeline, 131, 145
Harry, Ken, 167
Harry's Crossing, 56
Harry's Dam, 141
Harvie, Eric, 179
Harvie, Neil, 179–80
Harvie, Tim, 179–80
harvesting, 40, 122–28
Hatheume Lake, 57
health and safety, 128–30
Histed, John, 138
Holden, William, 174
Holland, Cuyler, 183–84
Holmes, Chief Fred (Scotty), 167
Holmes, Louis, 167
Home Ranch, 13, 32, 43–44, 56, 61, 76–77, 84, 106, 124, 130, 142–54
horses, 38, 52–54
Best Remuda Award, 53, 59, 61
Clydesdale horses, 21, 44
Quarter Horse, 38, 52–54, 77
housing, 54, 91, 142, 148, 151, 167
Huffman, Cuyler, 183–84
Huffman, Grant, 180, 182
Hurlburt, Brant, 106

I
irrigation, 24, 26–27, 41, 51, 88, 93, 100, 107, 122–24, 126–28
"Island" at English Bridge, 61

J
Jacobs, Cameron, 103, 157
Jacobs, Megan, 103, 156–57
Jacobs, Shirley, 156
Jacobs, Stan, 63–65, 162
Jacobs, Trina, 103
James Cattle Company, 91, 176
James, Dale, 175–76
James, Daryl, 175
James, Don, 175
James, Helen, 175
James, Janet, 175
James, Linda, 175
James, Lyle, 91–92, 175–76
James, Marilyn, 175
James, Mary, 91–92, 175–76
James, Susan, 175
James Cattle Company, 91, 176
Janzen, Dr. Eugene, 54
Johnson, Charles E. Wynn, 29
Jones, Shauna, 112
Junction Sheep Range Provincial Park, 181

K
Kanji, Al, 83
Kempston, Marv, 132
Kenver Equipment, 89
King Charles III, 38
Kingdon, Miles, 63
Koster, John, 172
Koster, Henry, 172
Kroenke, Stan, 87–88, 169, 171

L
Latta, Emily, 71
Latta, Mike, 58, 71
Lauder, Ian, 164
Lauder, Dr. Jean (Hansen), 164
Lauder, John, 164
Lauder, Joseph Dixon, 164
Lauriente, Kathy, 182
Lavictoire, Mary, 144
Lewis, Cathy, 67
Little Chapperon Lake, 141
Little Dog Ranch, 174
Little Sabin Lake, 141
Livingston, Tom, 180
Lock, Kathie, 144
logging, 132–34
Louis Corrals, 144

M
Magee, Hugh, 46, 50–51, 157
Magee, Sherri, 51
Manuel, Chief Herbie, 72
Martindale, Curt, 92
Martindale, Erica (Huber), 92
McCauley, Joe, 37
McDermid, Bill, 143–44
McDermid, Tereza, 143
McGill, Jim, 82
McKenzie, Dale, 52
McKenzie, Jerry, 52
McKenzie, Laura, 52
McKenzie, Linda, 52
McKenzie, Ryan, 52
McKimm, George, 179
McNee, Sir David, 74
McNee, Isabel, 74
McRae, Clifford, 167
McRae, Harold (Spook), 167
McRae, Wayne, 167
McVey, Peter, 72, 120
Meason, William Liang, 174
Mellin Lake, 141
Mergle, Sally, 144
Mervyn, Doug, 173
Mervyn, Marie, 173
M.G. Drummond Ranch, 179
Milliken, Terry, 52, 63, 143–44

Miner, Billy "Grey Fox" (George Edwards), 153
Minnie Lake, 56, 121, 126, 140–41, 144
Minnie Lake Ranch House, 141
Minor, John, 179
Mitchell, Austin, 71–72
Moon, Charlie, 179, 181
Moon, Jack, 181
Moon, Rob, 181
Mountain Ranch, 174
Munro, Joseph, 167
Murphy, Stan, 52, 65, 70, 81
Murray, Sheila, 61
Murray, Stewart, 61, 122, 124, 126–28, 145, 167

N
National Cattlemen's Beef Association, 70
Ned, Ron, 167
Nelson, Melvin, 187
Nicola Stock Breeders Association, 70
Nicola Ranch, 52, 96
Nicola Valley Fish and Game Club, 120–21
Nielsen, Arnold, 56, 61
Nielsen, Folmer, 150
Norfolk Ranch, 44, 61
northern operations, 97–100, 122, 124, 128, 130, 145, 173

O
O'Byrne, Christine, 44, 60, 63
O'Byrne, Mark, 63
O'Byrne, Tim, 60, 63
Occupational Health and Safety, 129
O'Reilly, Peter, 17, 19
Oughtred, Cliff, 84, 86
Oughtred, Margo, 84
Ownership Identification Inc., 70

P
Panorama Sale, 63, 70, 106
Paradise Valley, 172
Parkes, Bobbi, 7, 96
Parkes, Duke, 96
Parkes, John (JP), 96
Parkes, Will, 96
Parks, Brittney, 97
Patton, Thomas, 174
Patton, William, 174
Pemberton, Joseph Despard, 17
Pennask Lake, 57, 120, 166
Pennask Lake Fishing and Game Club, 166
Penny, Wes, 120, 129
"Peppy San", 38
Pigeon Meadows Ranch, 174
Pigeon, Moyse (Moses), 174
Pikes Lake, 141

Place, Joseph Smith, 174
Plummer, Trena, 181
Plummer, Wayne, 181
Pockrant, Leah, 149–50
Portland cow camp, 33, 64
Prentice, James Douglas, 184
Prince Philip, Duke of Edinburgh, 38, 39

Q
Quilchena Cattle Co., 92, 178
Quilchena general store, 106, 141
Quilchena Hotel, 52, 92, 140–41, 177
Quilchena Motors, 178
Quilchena Ranch, 177–78, 92–93
Quilchena Creek long-stay RV facility, 141
Quilchena RV Park and Marina, 141

R
Ramstad, Bev, 187
Ramstad, Larry, 63, 187
Reese, Dick, 50–51
Reese, Myrtle, 50–51
Reimer, Bill, 44
Reimer, Christine, 44
Reimer, Fred, 44
Reimer, Joan, 44
Reimer, John, 44
Richmond, Claude, 74
Riske Creek, 93, 112, 122, 124, 130, 179–83
Riske, L.W., 179
Robinson, Wayne, 186
Rose, Anne, 178
Rose, Guy, 178
Rose, Hilde, 178
Rose, Matlock, 53
Rose, Mike, 178
Rose, Paul, 178
Rose, Peter, 178
Rose, Steve, 178
Ross, Frank MacKenzie, 26–27, 186
Roulston, Candice, 57
Roulston, Orval, 57–58, 63
Round Lake (Sanctuary Lake), 15, 162
Royal Windsor Horse Show, 38–39
Rudiger, John, 187

S
Sabin Lake campsites, 141
Sageishi, Yuki, 138
Salmon Lake (Fish Lake), 36–37, 44, 84, 135, 137–39, 141, 148–49, 166
Salmon Lake Resort, 141
Sanctuary Lake (Round Lake), 15, 162
Sanders Contracting, 80
Sanders, Jerry, 80–81
Sanders, Ron, 81

Sauvé, Governor General Madame Jeanne, 58
Sawmill Lake (Rush Lake), 51
Schnurrenberger, John, 54, 153, 154–55
Schreyer, Governor General Edward, 72
Schreyer, Lilly, 72
Seasonal Agricultural Worker Program, 100
Sidwell, Melvin, 186
Simpson, Jeff, 131
Simpson Sawmills, 131
Skelton, Floyd, 185
Smith, Amanda, 148
Smith, Cody, 148
Smith, Keith, 148
Smith, Sandy, 148
Snow Walker, The, 137
South Alleyne Lake, 141
Spencer, Barbara, 174
Spencer, Victor, 26, 168–69, 174, 186
Springbank Ranch, 164
Stein, Gordon, 72
Stein, Norm, 57
Stewart, Don, 145
Stewart, Louie, 167
Stewart, Martin, 167
Stewart, Mary, 145
Stewart, Nancy, 167
Stewart-Brewer, Teresa, 145, 148
Stoltzfus, Leann, 59
Stoltzfus, Madison, 59
Stoltzfus, Reesa, 59
Stoltzfus, Wendell (Puck), 59, 61
Stoney Lake, 120–21, 130, 135
Stoney Lake Lodge, 80, 84, 137, 139, 141, 149
Stringam, Bryce, 179
Studdert, W.P. (Bill), 26, 168, 185–86
Sun Cattle Co., 174

T
The Thaw, 176
Thibeault, Bernice, 66–67
Thibeault, Casey, 66–67
Thibeault, Clay, 66–67
Thibeault, Trevor, 63, 66–67, 166
Thomson, Charles William Ringler, 17
Tidball, George, 92
Tidball, Steve, 92
Triangle Ranch, 177
Turner, Jim, 172
Turner, John, 27
Turner, Phyllis, 27
Twan, Bill, 173
Twan, Bronc, 173
Twigg's Place, 36, 40, 84, 137, 139, 141
Two Rivers Ranching, 180–81
Tyson, Ian, 76, 160

U

University of Saskatchewan, Western College of Veterinary Medicine, 54
Upper Nicola Band, 72, 82, 130, 167

V

Valenzuela, Raphael, 174
Valley Auction, 106
Van Volkenburgh, Benjamin, 17, 19
Verified Beef Production Plus, 79
Versepuche, Le Comte de Isadore (aka Gaspard), 174
VJV Auctions, 106
Vokel, Dick, 71
Vold, Blair, 106
Von Riedemann, Mario, 173
Von Reidemann, Martin, 173

W

Wade, John, 179
Ward, Frank Bulkley, 23–25
Ward, William Curtis, 17–18, 23–24
Wasley Lake, 141
West, Ann, 32
West, John Joseph, 29–31, 38, 40
West, John Jr., 33–34, 38

West, June, 32
West Bay Construction, 84
Western Canadian Ranching Company (WCRC), 184
Westwood Fibre Resources, 132
Whipple, Lynne, 152
White, Mary "Twigg", 36–37, 137
White, Melanie, 36
White, Rhegan, 36–37
WhiteLine Road Maintenance, 81
wildlife species, 112–117
Winch, Gertrude, 26
Woodman, Carl, 80
Woodman, Rebecca, 80
Woodward, Carol, 29, 45–46, 71–72, 25, 84, 156
Woodward, Charles A., 29
Woodward, Charles Namby Wynn "Chunky", 27, 29–30, 36–40, 45–46, 58, 67, 71–73, 75–76
Woodward, John, 38, 76, 81–82
Woodward, Kip, 7, 32–35, 50, 75–76, 77, 80–82, 84
Woodward, Robyn, 34–35
Woodward, William C., 29
Woodward, Wynn, 32

Woodward's Food Floors, 105
Woodward's Stores Limited, 29, 43
Woolliams, Neil, 40–41, 45, 169, 173
Workers' Compensation Board (WCB), 128
Workers' Compensation Bravery Award, 58
Working Ranch Magazine, 63
WorkSafeBC, 128
WorldCom, 84–86
Wright, Bill, 174

Y

Young, John, 57–59